Optical Interference and Dynamic Diffraction

Research methods for undergraduates

Online at: https://doi.org/10.1088/978-0-7503-4836-2

IOP Series in Advances in Optics, Photonics and Optoelectronics

SERIES EDITOR

Professor Rajpal S Sirohi Consultant Scientist

About the Editor

Rajpal S Sirohi is currently working as a faculty member in the Department of Physics, Alabama A&M University, Huntsville, AL, USA. Prior to this, he was a consultant scientist at the Indian Institute of Science, Bangalore, and before that he was Chair Professor in the Department of Physics, Tezpur University, Assam. During 2000–2011, he was an academic administrator, being vice-chancellor to a couple of universities and the director of the Indian Institute of Technology, Delhi. He is the recipient of many international and national awards and the author of more than 400 papers. Dr Sirohi is involved with research concerning optical metrology, optical instrumentation, holography, and the speckle phenomena.

About the series

Optics, photonics, and optoelectronics are enabling technologies in many branches of science, engineering, medicine, and agriculture. These technologies have reshaped our outlook and our ways of interacting with each other, and have brought people closer together. They help us to understand many phenomena better and provide deeper insight into the functioning of nature. Further, these technologies themselves are evolving at a rapid rate. Their applications encompass very large spatial scales, from nanometres to the astronomical scale, and a very large temporal range, from picoseconds to billions of years. This series on advances in optics, photonics, and optoelectronics aims to cover topics that are of interest to both academia and industry. Some of the topics to be covered by the books in this series include biophotonics and medical imaging, devices, electromagnetics, fibre optics, information storage, instrumentation, light sources, charge-coupled devices (CCDs) and complementary metal oxide semiconductor (CMOS) imagers, metamaterials, optical metrology, optical networks, photovoltaics, free-form optics and its evaluation, singular optics, cryptography, and sensors.

About IOP ebooks

The authors are encouraged to take advantage of the features made possible by electronic publication to enhance the reader experience through the use of color, animation, and video and by incorporating supplementary files in their work.

A list of recent titles published in this series can be found here: https://iopscience.iop.org/bookListInfo/series-on-advances-in-optics-photonics-and-optoelectronics.

Optical Interference and Dynamic Diffraction

Research methods for undergraduates

Jenny Magnes

Physics and Astronomy Department, Vassar College, Poughkeepsie, NY, USA

Juan M Merlo-Ramírez

Physics and Astronomy Department, Vassar College, Poughkeepsie, NY, USA

IOP Publishing, Bristol, UK

Contents

Preface

Light is one of the most studied and used physical phenomena in Nature. It spans classical physics as well as quantum mechanics. In this text, we focus on the classical phenomenon of diffraction with the goal of understanding **dynamic optical diffraction (DOD)**. Chapters 1–5 set the stage by examining static optical diffraction using the appropriate mathematical tools such as Fourier transforms The physics and mathematics in this text are at the undergraduate level. The authors aimed to detail the mathematics in ways that facilitate undergraduate development. Chapters 6 and 7 dive into DOD and its applications. The appendices include explanations on some common optical devices used in conjunction with diffraction, safety tips, and solutions to practice problems.

After years in classrooms and laboratories, diffraction still strikes us as more than a set of pretty fringes. What we see in diffraction are details beyond unaided vision, motion where cameras blur, and patterns that hint at order, nonlinearity, and sometimes chaos.

The aim of this book is to make that vision practical. We show how DOD transforms evolving interference patterns into quantitative and analyzable signals rich in information, and we provide a clear framework for using those signals to understand physical and biological systems. Rather than treating time as a nuisance, we place it at the center: change itself becomes the measurement.

The book proceeds deliberately. The first five chapters establish the optical and mathematical foundation needed to read diffraction patterns with confidence: interference, wave equations, Fourier optics, and computational diffraction. Chapters 6 and 7 make the transition to DOD and its applications, demonstrating how time-resolved diffraction encodes frequency content, nonlinear structure, and dynamical markers. The appendices collect the practical tools: sources and detectors, essential optical hardware, and safety practices, so that readers can implement the experiments with care.

Our hope is that these pages help you look past the fringes and see the dynamics within.

Acknowledgements

Any substantial body of work is rarely the product of a single individual. Even when only one or two names appear on the title page, progress depends on the dedication of many people and the support of complex systems that make research possible.

We are deeply grateful to Vassar College for fostering an environment in which new academic directions can be explored, established ideas can be further developed, and collaborations can emerge organically to expand knowledge. This spirit of openness and intellectual exchange has been essential to our work. The Undergraduate Research Summer Institute (URSI) has been a particularly valuable source of collaboration with undergraduates, sustaining innovation and curiosity in our group. We also thank the support staff, especially Christine Anzalone, whose attention to equipment and supplies ensured that our experiments could run smoothly.

The development of DOD has been shaped by the generosity and insight of many colleagues. We thank Professor Kathleen Raley-Susman for providing live *C. elegans* samples and for teaching us how to cultivate and study these organisms, a time-intensive work that grounded our efforts to connect biology with optical physics. Professor Jianwei Miao inspired our exploration of diffraction through oversampling and image reconstruction, and generously shared his reconstruction code. Dr Susannah Zhang played an instrumental role in validating DOD; her doctoral work relied on this method, and her positive spirit lifted our group while her rigor kept us focused. We also honor Professor Harold Hastings, whose encouragement, humor, and vision remain a lasting presence in our research.

Finally, we acknowledge with gratitude the steady encouragement of our families and friends throughout long days and weekends in the lab. Professor Jenny Magnes dedicates her work to her children, Sophie and Sam, and is especially grateful for the unwavering support of her best friend, Mary Jikhars, and her dear friend, Yolande Goodman. Profesor Juan M Merlo-Ramírez dedicates his work to his children, Juan and Ximena, for their patience during long hours of research and especially to his wife, Ana, whose care and support made it possible for him to remain fully devoted to this project.

Author biographies

Jenny Magnes

Jenny Magnes is a Professor of Physics in the Department of Physics and Astronomy at Vassar College. She earned a BS in English from the University of Maryland (European Division), a BS in Physics and Mathematics from Delaware State University, and a Masters degree in Physics from Temple University. In 2003, she earned her PhD in physics from Temple University in Philadelphia with a dissertation on high-resolution diatomic spectroscopy. Her current research is at the intersection of optics, neuroscience, and deterministic chaos. As a veteran of the US Army, she has advocated for and mentored veterans returning to college.

Juan M Merlo-Ramírez

Juan M Merlo-Ramírez is an Associate Professor in the Department of Physics and Astronomy at Vassar College. He holds a five-year degree in Physics and an MSc in Optoelectronics from the University of Puebla. In 2010, he earned a PhD in Optics from the National Institute of Astrophysics, Optics, and Electronics (INAOE) in Puebla, Mexico, with a dissertation on near-field microscopy. His current research focuses on two main areas: (1) near-field microscopy and plasmonics, where he studies light–matter interactions at the nanoscale, and (2) topological phases of matter in classical systems, with an emphasis on photonic and mechanical topological insulators. He is also interested in disseminating scientific knowledge by writing science books for children.

IOP Publishing

Optical Interference and Dynamic Diffraction
Research methods for undergraduates
Jenny Magnes and Juan M Merlo-Ramírez

Chapter 1

Introduction

1.1 Introduction

When light bends around an obstacle, scatters from a surface, or passes through an aperture, it leaves telltale patterns. These patterns, the fingerprints of diffraction and interference, have been central to physics for centuries. They revealed the wave nature of light in the nineteenth century, helped crack the double helix structure of DNA in the twentieth century, and remain indispensable across science and engineering today.

Traditionally, diffraction has been studied under static conditions: a fixed aperture, a crystal at rest, or a still medium. The resulting patterns are also static, representing snapshots frozen in time. But what if the object itself is in motion? What if it bends, squirms, or evolves while illuminated? In such cases, the diffraction pattern is no longer fixed. Fluctuating dynamically encodes not only the structure of the object but also its motion.

This is the essence of dynamic object diffraction (DOD). By shining a coherent light source such as a helium neon laser onto a living or moving sample, we obtain diffraction patterns that evolve in real time. Recording intensity fluctuations at selected points in these patterns generates time series that capture the global dynamics of the system. From these signals, oscillation frequencies, harmonics, and even indicators of nonlinear behavior such as chaos can be extracted.

1.2 Light, interference, and diffraction

According to the Huygens–Fresnel principle, each point on an aperture acts as a source of secondary spherical waves. In the far-field (Fraunhofer) limit, the superposition of these waves yields an intensity pattern that is the squared magnitude of the Fourier transform of the aperture function (diffracting object).

Obstacles change the shape of a wavefront, leading to a diffraction pattern that can be observed after the light passes the obstacle. Different types of diffraction

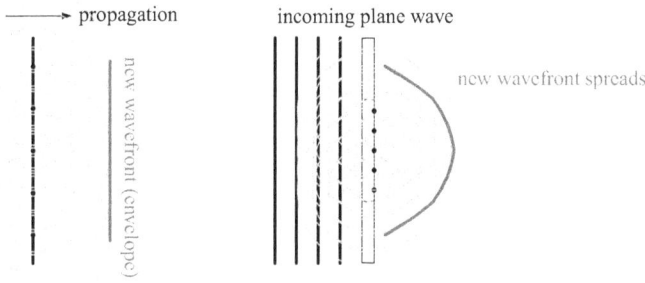

Figure 1.1. (Left) Plane wave without obstacle continues to propagate with a uniform wavefront. (Right) A plane wave encounters an obstacle and spreads after the obstacle.

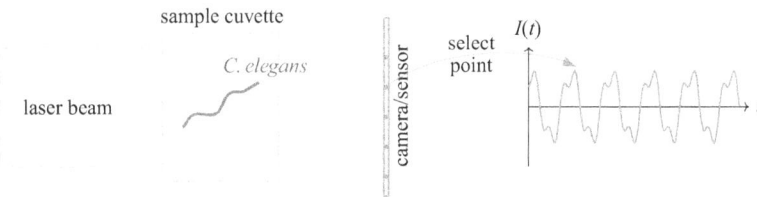

Figure 1.2. DOD concept: a coherent light beam oversamples a moving organism. The evolving diffraction pattern is recorded, and a fixed point yields a time series $I(t)$.

patterns can be observed near or far from the obstacle. In this text, we focus on far-field diffraction, which is also called Fraunhofer diffraction (figure 1.1).

1.3 From static to dynamic diffraction

If the aperture or diffracting object changes with time, for example, an organism undulates as a wavefront encounters it, the diffraction pattern will also be time dependent. We wonder how the diffraction pattern relates to the changing/moving diffracting object. What is the relationship between the fluctuating intensity $I(t)$ at a fixed point in the diffraction pattern and the aperture function (diffracting object). Although diffraction at a single point compresses spatial detail, $I(t)$ integrates over the entire object, yielding a compact yet powerful record of global dynamics.

DOD encodes the relative spatial dynamic in a one-dimensional time series. This series provides information about frequencies and complexity (figure 1.2).

1.4 Frequency analysis of motion

Because the signal $I(t)$ is time dependent. Frequency analysis indicates characteristic locomotion frequencies and harmonics. In nematodes, for example, a fundamental peak near ~ 1 Hz often reflects the undulation rate; even harmonics can be enhanced by intensity detection and body shape symmetries, while odd harmonics signal waveform asymmetries and nonlinearity.

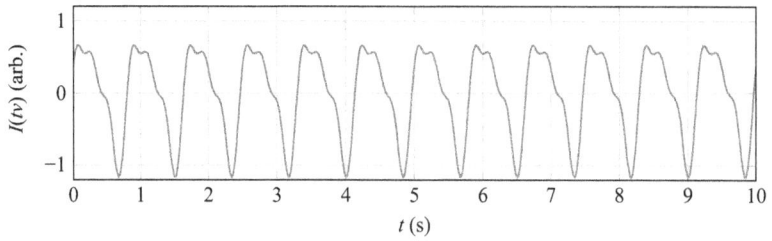

(a) Example intensity time series $I(t)$ with fundamental and harmonics.

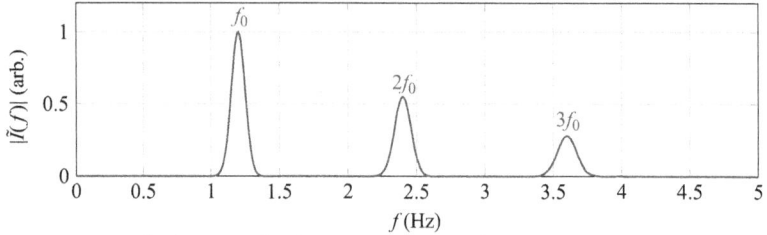

(b) Corresponding frequency spectrum with peaks at f_0, $2f_0$, $3f_0$

Figure 1.3. Fourier analysis reveals characteristic motion frequencies and harmonics.

The frequency spectrum displays locomotory frequencies across several length scales with great precision. These types of spectra are an initial indicator for locomotary dynamics and complexity (figure 1.3).

1.5 Beyond linear oscillations: chaos and complexity

DOD time series also lend themselves to nonlinear analysis. Using delay-coordinate embedding (Takens' theorem), one reconstructs a state-space dynamics from a single observable.

$$\mathbf{x}_n = (I(n\Delta t),\, I((n + \tau)\Delta t),\, \ldots,\, I((n + (m - 1))\tau\Delta t)), \qquad (1.1)$$

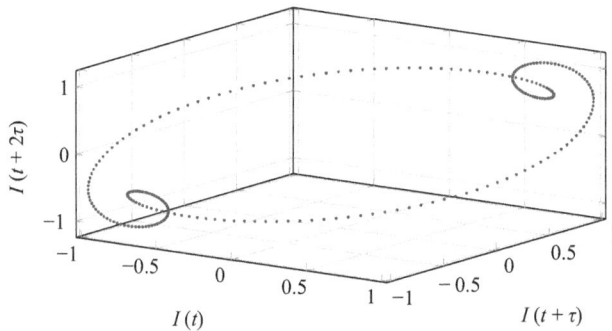

Figure 1.4. Delay-embedding of a synthetic DOD signal (embedding dimension $m = 3$, delay τ chosen for demonstration). The folded structure hints at quasiperiodicity/nonlinearity; real data may show thicker filaments and positive largest Lyapunov exponent (LLE).

where n is a positive integer indicating the nth state-vector $\mathbf{x}_n$. $t_n = n\Delta t$ and $\tau_{time} = \tau \, \Delta t$, where Δt is the sampling interval and τ the time delay to reconstruct the dynamics in phase space. m represents the embedding dimension equally the number of components in each vector $\mathbf{x}_n$

If nearby phase trajectories diverge on average, the LLE is positive, is a marker for deterministic chaos (figure 1.4).

1.6 Why dynamic optical diffraction matters

Dynamic diffraction is more than a measurement technique; it is a perspective. By treating diffraction patterns as evolving records of motion, the DOD opens new avenues for studying biological and physical systems. For biology, it complements traditional microscopy and video analysis. While video provides local body-shaped information, the DOD condenses global motion into analyzable time series. In physics, it can probe fluid instabilities, colloidal dynamics, or structural phase transitions.

Ultimately, by shifting the view from static patterns to dynamic signatures, DOD gives access to hidden rhythms in complex systems. The chapters that follow develop the optical and mathematical foundations, extend them to dynamic settings, and demonstrate practical implementations and applications across biology and beyond.

1.7 Summary

This text presents the fundamental physical principles behind diffraction with the goal of formally introducing DOD. These physical principles include fundamentals of electromagnetic radiation, interference, and diffraction. We introduce mathematical tools such as Fourier transforms in a way that allows the mathematical and physical intuition to connect.

The science and mathematics introduced here are meant to be targeted rather than exhaustive. Explanations and visuals are designed to help upper-level undergraduate students understand the components of the principles behind diffraction so that they can extend that knowledge to the time-dependent case. We refer to specialized texts for other diffraction methods and a deeper understanding of Fourier analysis.

The last chapter touches briefly on the analysis of DOD time series; however, for analysis of chaotic systems such as the one presented, we refer the reader to specialized texts [1, 2].

References

[1] James J F 2011 *A Studentas Guide to Fourier Transforms: With Applications in Physics and Engineering* (Cambridge: Cambridge University Press)
[2] Brigham E O 1988 *The Fast Fourier Transform and Its Applications* (Hoboken, NJ: Prentice-Hall)

IOP Publishing

Optical Interference and Dynamic Diffraction
Research methods for undergraduates

Jenny Magnes and Juan M Merlo-Ramírez

Chapter 2

Optical interference

2.1 Introduction

This chapter is dedicated to the study of wave interference, its origin and properties. We begin by outlining the basic properties of waves and deriving the one-dimensional wave equation. Then, we find and analyze the solutions that describe standing and traveling waves in open space. The treatment is then extended to higher dimensions introducing the idea of wavefront.

Next, we focus on the principle of superposition and show the interaction between multiple waves one and two dimensions. We analyze how phase shifts and wave vector orientations determine regions of constructive and destructive interference.

The chapter concludes by linking these concepts to electromagnetic waves, deriving their governing equations from Maxwell's laws and discussing the role of polarization in interference patterns.

2.2 Fundamental definition of a wave

The concept of waves is fundamental for the understanding of core concepts in physics. From the gentle ripples on a pond to the oscillating fields of light and the probabilistic vibrations of matter in quantum mechanics, waves represent the propagation of disturbances in space and time. A wave is fundamentally a mechanism for the propagation of energy and information. Unlike the transport of matter, where particles move, a wave transmits energy through oscillations or disturbances of a medium without any net movement.

2.3 Mathematical description of a wave

Despite the wide range of wave phenomena we could find in Nature, a wave can be described as a perturbation that propagates through space and time. It is important to remark that waves only allow the propagation of energy. This means that we can describe a wave independently of the medium in which it propagates, only

doi:10.1088/978-0-7503-4836-2ch2

depending on the space and time. As such, a one-dimensional wave is described by a function of position, x, and time, t, as:

$$\Psi(x, t) = f(x - v\,t), \tag{2.1}$$

where $\Psi(x, t)$ is the so-called wave function, v is the wave velocity, and f is any function describing the shape of the wave.

2.4 Derivation of the wave equation

We have seen that the general representation of a wave is given by equation (2.1). However, it is useful to determine what kind of functions f can be used. The wave equation is a way to find suitable functions for f. To derive the wave equation, we need to assume that $\Psi(x, t)$ is continuous and differentiable at least up to its second derivative. Then, we determine the change of $\Psi(x, t)$ along the space coordinate x as:

$$\frac{\partial \Psi(x, t)}{\partial x} = f'(x - v\,t), \tag{2.2}$$

Taking the second derivative, we get:

$$\frac{\partial^2 \Psi(x, t)}{\partial x^2} = f''(x - v\,t). \tag{2.3}$$

Similarly, we find the way $\Psi(x, t)$ changes over time t up to its second derivative:

$$\frac{\partial \Psi(x, t)}{\partial t} = -v\,f'(x - v\,t), \tag{2.4}$$

Taking the second derivative:

$$\frac{\partial^2 \Psi(x, t)}{\partial t^2} = v^2 f''(x - v\,t). \tag{2.5}$$

We can see that equations (2.3) and (2.5) have $f''(x - v\,t)$ on their right side, up to a factor v^2, then it is easy to see the following:

$$\frac{\partial^2 \Psi(x, t)}{\partial t^2} = v^2 \frac{\partial^2 \Psi(x, t)}{\partial x^2}, \tag{2.6}$$

Rearranging equation (2.6), we get the **one-dimensional wave equation** as:

$$\frac{\partial^2 \Psi(x, t)}{\partial x^2} = \frac{1}{v^2} \frac{\partial^2 \Psi(x, t)}{\partial t^2}. \tag{2.7}$$

It is clear from equation (2.7) that the one-dimensional wave equation does not depend on the phenomenon generating the wave. Moreover, this equation can be employed in fields as different as mechanics, acoustics, and electromagnetic waves using the appropriate parameters.

2.5 Generalization to a three-dimensional wave equation

Although in terms of notation, the change from one to three dimensions is quite simple, there are profound implications in the meaning of the wave. For example, we have defined the wave as $\Psi(x, t)$ for one dimension, x being the direction of propagation. In the three-dimensional case, we now use $\mathbf{r}$[1] for the position, which results in $\Psi(\mathbf{r}, t)$.

Now, if we want to determine the spatial second derivative in a three-dimensional space, as in equation (2.3), we have to use the Laplacian operator ∇^2 [1][2]. In this way, the three-dimensional wave equation takes the form:

$$\nabla^2\Psi(\mathbf{r}, t) = \frac{1}{v^2}\frac{\partial^2\Psi(\mathbf{r}, t)}{\partial t^2}. \tag{2.8}$$

As the operator ∇ depends on the coordinate system used, it can take quite different forms The reader is advised to refer to chapter 7 of [1] for a better understanding of this point.

2.6 Explicit forms of $\Psi(x, t)$

When we talk about waves, we usually think of periodic perturbations of space and time, as we mentioned before. In this sense, there are several ways to represent actual waves by solving the wave equation, equation (2.8). So, any function $\Psi(x, t)$ that satisfies equation (2.8) can be referred to as a wave.

In order to show a simple example, the next section is dedicated to finding a solution to the one-dimensional wave equation (equation (2.7)).

2.6.1 One-dimensional wave

Standing wave

The one-dimensional wave equation is a second-order linear partial differential equation [2]. As such, we can find the solution to equation (2.7) using any of the well-known methods in a differential equations textbook [1]. However, a quite simple method is the separation of variables that will be used later in this book, so we will summarize its most important steps.

We consider the wave equation:

$$\frac{\partial^2\Psi(x, t)}{\partial t^2} = v^2\frac{\partial^2\Psi(x, t)}{\partial x^2}, \tag{2.9}$$

with

$$0 < x < L, \quad t > 0, \tag{2.10}$$

---

[1] Throughout the rest of this textbook, we will use bold letters to denote vectors.
[2] The Laplacian operator, in Cartesian coordinates, can be defined as $\nabla^2 = \frac{\partial^2}{\partial x^2} + \frac{\partial^2}{\partial y^2} + \frac{\partial^2}{\partial z^2}$. However, its definition depends on the coordinate system used.

where L is the spatial domain under analysis. The boundary conditions are assumed to be:

$$\Psi(0, t) = 0, \quad \Psi(L, t) = 0, \tag{2.11}$$

and initial conditions:

$$\Psi(x, 0) = f(x), \quad \left.\frac{\partial \Psi(x, t)}{\partial t}\right|_{t=0} = g(x). \tag{2.12}$$

Using the separation of variables, we assume a solution of the form:

$$\Psi(x, t) = X(x)T(t). \tag{2.13}$$

with $X(x)$ and $T(t)$ functions depending on the spatial coordinate x and the time t, respectively.

Substituting into equation (2.7) results in:

$$X(x)T''(t) = v^2 X''(x)T(t), \tag{2.14}$$

which leads to:

$$\frac{T''(t)}{v^2 T(t)} = \frac{X''(x)}{X(x)} = -\lambda. \tag{2.15}$$

with λ a factor related to the eigenvalues of the solutions $X(x)$ and $T(t)$.

It is clear that equation (2.15) can be written as two ordinary differential equations:

$$X''(x) + \lambda X(x) = 0, \tag{2.16}$$

and

$$T''(t) + \lambda v^2 T(t) = 0. \tag{2.17}$$

It is important to remember that equation (2.16) has boundary conditions equivalent to equation (2.11) as:

$$X(0) = 0, \quad X(L) = 0. \tag{2.18}$$

Solving the spatial equation, equation (2.16), we get the eigenfunction:

$$X_n(x) = \sin\left(\frac{n\pi x}{L}\right), \quad n = 1, 2, 3, \ldots \tag{2.19}$$

In the same fashion, the solution to the time equation, equation (2.17), results:

$$T_n(t) = A_n \cos\left(\frac{n\pi v\, t}{L}\right) + B_n \sin\left(\frac{n\pi v\, t}{L}\right). \tag{2.20}$$

We can find the general solution by combining equations (2.19) and (2.20) following equation (2.13), to get:

$$\Psi(x, t) = \sum_{n=1}^{\infty}\left[A_n \cos\left(\frac{n\pi v\, t}{L}\right) + B_n \sin\left(\frac{n\pi v\, t}{L}\right)\right] \sin\left(\frac{n\pi x}{L}\right). \tag{2.21}$$

Equation (2.21) represents a set of modes that allow our system to oscillate, also called *normal modes*, i.e., the wavy characteristic of a wave.

Although equation (2.21) is a complete solution, it is useful to know the value of the coefficients A_n and B_n. We can find the value of A_n using the initial condition stated in equation (2.10) as:

$$\Psi(x, 0) = f(x) = \sum_{n=1}^{\infty} A_n \sin\left(\frac{n\pi x}{L}\right), \tag{2.22}$$

This means that:

$$A_n = \frac{2}{L} \int_0^L f(x)\sin\left(\frac{n\pi x}{L}\right) \mathrm{d}x. \tag{2.23}$$

Determining the value of B_n is similar to the previous procedure, but this time using the initial velocity in equation (2.11) as follows:

$$\left.\frac{\partial \Psi(x, t)}{\partial t}\right|_{t=0} = g(x) = \sum_{n=1}^{\infty} \frac{n\pi v}{L} B_n \sin\left(\frac{n\pi x}{L}\right). \tag{2.24}$$

This results in:

$$B_n = \frac{2}{n\pi v} \int_0^L g(x)\sin\left(\frac{n\pi x}{L}\right)\mathrm{d}x. \tag{2.25}$$

It is important to note that the procedure that we just used to solve the one-dimensional wave equation is based on the fact that the system under analysis is bounded, i.e. it is limited to the region $0 < x < L$. This means that our general solution, equation (2.21), is a standing wave; it is a wave that does not propagate [2]. In fact, this system can be found in several areas of physics, such as acoustics, mechanics, electromagnetic theory, and even quantum mechanics.

A very instructive example of the use of the solution in equation (2.21) is a string fixed at one end and excited at the opposite end. When the excitation frequency matches an integer of π/L, with L being the length of the string, the normal modes are expressed. Figure 2.1 shows the first six normal modes in an elastic string at its resonant frequencies.

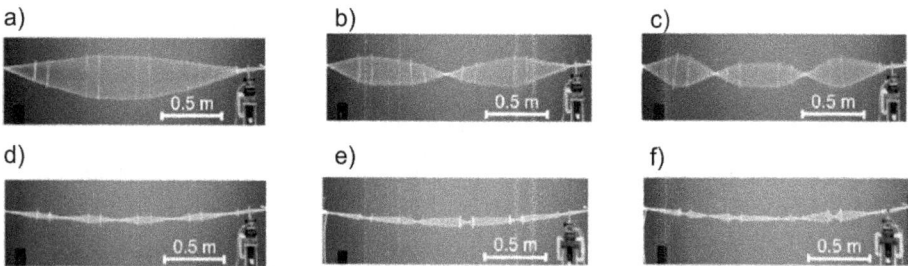

Figure 2.1. First six resonant modes in a bounded elastic string. Credit: Merlo Lab 2022.

Propagating wave

Although the solution shown in equation (2.20) is very useful, we still want to cover other important case, the so-called *propagating wave*. In such a case, the wave is not confined to a bounded region of space. Thus, the boundary conditions in equation (2.11) do not apply.

In order to find a solution to the one-dimensional wave equation, equation (2.7), we make use of the d'Alembert method. This method consists in assuming that the solution is a linear superposition of traveling waves as:

$$\Psi(x, t) = f(x - v t) + f(x + v t). \tag{2.26}$$

Using the same initial conditions as in equation (2.12), the general solution using the D'Alembert method is:

$$\Psi(x, t) = \frac{1}{2}[f(x - v t) + f(x + v t)] + \frac{1}{2v} \int_{x-v t}^{x+v t} g(s) \, \mathrm{d}s. \tag{2.27}$$

In order to find a suitable solution, we can take a look at equation (2.21). There, we see that the solution is expressed in terms of sine and cosine functions. As such, if we select any combination of these functions in equation (2.27), we could get a stationary wave due to the counter-propagating nature of the solution. However, if we arbitrarily select only one direction and $g(x) = 0$, equation (2.27) becomes a propagating wave. Then, we can assume a particular solution as follows:

$$\Psi(x, t) = \sum_{n=1}^{\infty}[A_n \sin(k_n x - k_n v t) + B_n \cos(k_n x - k_n v t)], \tag{2.28}$$

with k_n the wave number, that physically represents the spatial frequency of the wave. This term will become important in chapter 3, when we discuss the diffraction of electromagnetic waves.

In addition, equation (2.28) shows the relationship $k_n v$, that is, the product of the wave number and the velocity of the wave, which is also known as the *dispersion* of the wave, defined as $\omega = k_n v$, with ω_n the angular frequency. Then, equation (2.28) can be written as:

$$\Psi(x, t) = \sum_{n=1}^{\infty}[A_n \sin(k_n x - \omega_n t) + B_n \cos(k_n x - \omega_n t)] \tag{2.29}$$

An experimented reader would understand that equation (2.29) is a solution in terms of the Fourier series [3]. The values of A_n and B_n can be determined by the specific conditions of each case analyzed.

As we suggested at the beginning of this section, the solution of the wave equation can be a superposition of functions, however, a single term of equation (2.29) could also be a suitable solution. This means that the simplest mathematical representation of a propagating wave can be expressed as:

$$\Psi(x, t) = A_0 \cos(kx - \omega t + \phi), \tag{2.30}$$

with, k and ω the wave number and angular frequency, respectively, and A_0 the amplitude. We have added the phase ϕ in order to have a complete solution. The phase ϕ will be of interest when we study the interference of waves in the next section. It is important to note that the wave expressed by equation (2.30) has only one component with a single wave number and a single angular frequency. This kind of wave is called a *monochromatic wave*; contrary to that expressed in equation (2.30) that presents an infinite number of components or a *spectrum* of waves.

Up to this point, we have expressed the waves in terms of the wave number k and the angular frequency ω. We also mentioned the dispersion of the wave as:

$$\omega = kv. \tag{2.31}$$

We also said that k is the spatial frequency of the wave, which means that it is related to the extent of the wave in space. Actually, we can express this space extension as the *wavelength*, λ, as:

$$k = \frac{2\pi}{\lambda}. \tag{2.32}$$

Similarly, we can express the angular frequency in terms of the temporal frequency of the wave as $\omega = 2\pi\nu$. With the new definitions of the wavelength and temporal frequency, we can easily see that the dispersion, equation (2.31), can be rearranged as:

$$v = \lambda\nu. \tag{2.33}$$

Equation (2.33) is an important property of any wave, no matter its nature, relating the propagation velocity, also known as *phase velocity*, the wavelength, and the frequency.

Figure 2.2 shows a simple representation of a wave with the shape of a Gaussian pulse. We would like to note that the position of the pulse changes as time passes. This is consistent with our original definition of $\Psi(x, t)$. In this particular example, the shape of the pulse does not change along the propagation of the wave. This is due

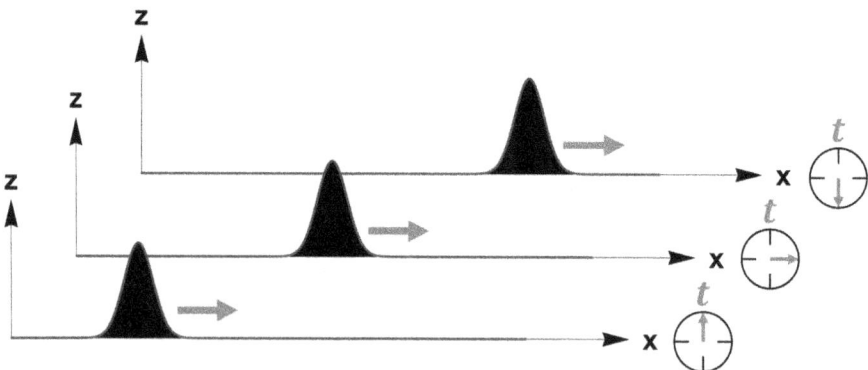

Figure 2.2. Propagating wave with the shape of a pulse at three different times.

to the lack of dispersion in the system. However, most of the actual physics experiments suffer from some degree of perturbation due to absorption in the propagating medium.

It is also possible for the reader to feel confused by the shape of the wave shown in figure 2.2, since it is not a sine or cosine function. However, it is important to remember that equation (2.29) can reproduce any function by finding the appropriate coefficients A_n and B_n.

In a more general representation, the use of the Euler's theorem [1] is commonly applied for the representation of the wave represented by equation (2.30). In this context, the wave can be written as:

$$\Psi(x, t) = A_0 \, e^{-i(kx - \omega t + \phi)}, \tag{2.34}$$

with i the imaginary unit.

Although equation (2.34) is a quite useful mathematical tool, it is important to remember that the magnitude of $\Psi(x, t)$, represented by the real part of equation (2.34), has a physical meaning, i.e. the amount of energy deposited by the wave in a certain area per unit of time; see appendix A for a better understanding.

2.6.2 Three-dimensional wave

Up to this point, we have described the perturbation $\Psi(x, t)$ as a one-dimensional entity. However, our Universe is a three-dimensional space, so a more general representation of a wave must be adopted. In this sense, a three-dimensional wave can be expressed as $\Psi(\mathbf{r}, t)$, with $\mathbf{r}$ a position vector[3]. In the three-dimensional case, the waves described by equations (2.21), (2.29), and (2.34) are still valid; however, we need to make the appropriate changes to the notation. In general, we must use three-dimensional versions of the position, $\mathbf{r}$, and the wave number, now called the *wave vector* $\mathbf{k}$.

When talking about three-dimensional waves, there are several shapes that are commonly used. Among the most common are the **plane**, **cylindrical**, and **spherical** waves. In fact, these are the waves that we will use throughout the rest of this textbook.

A *plane wave* is represented as:

$$\Psi(\mathbf{r}, t) = A_0 \, e^{-i(\mathbf{k} \cdot \mathbf{r} - \omega t + \phi)}, \tag{2.35}$$

Note that $\mathbf{k} \cdot \mathbf{r}$ actually represents a geometrical plane [2]. Figure 2.3(a) shows a representation of a plane wave; there, the direction of propagation, represented by the red arrow in figure 2.3, is perpendicular to the plane represented by $\mathbf{k} \cdot \mathbf{r}$.

Cylindrical and *spherical* waves can be represented using equations (2.36) and (2.37), respectively.

$$\Psi(\mathbf{r}, t) = \frac{A_0}{\sqrt{r}} \, e^{-i(kr - \omega t + \phi)}, \tag{2.36}$$

---

[3] As we mentioned before, we use bold letters to denote vectors.

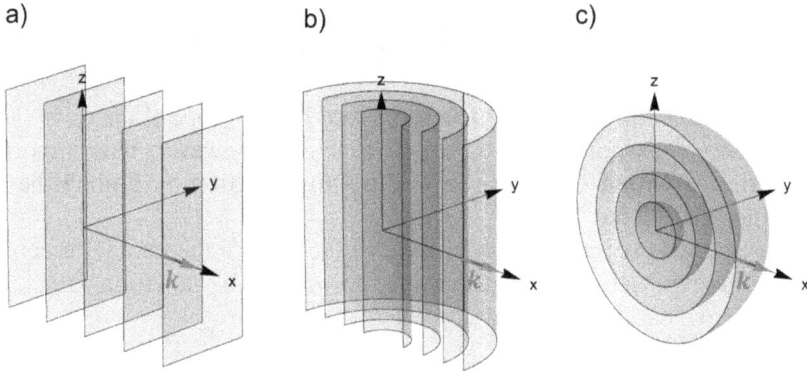

Figure 2.3. Schematic representation of a plane wave (a), a cylindrical wave (b) and a spherical wave (c). In all the cases, the red arrow represents the wave vector normal to the wavefront.

$$\Psi(\mathbf{r},\, t) = \frac{A_0}{r}\, e^{-i(kr - \omega t + \phi)}, \tag{2.37}$$

with $r = |\mathbf{r}|$.

Figures 2.3(b) and (c) show a representation of a cylindrical and a spherical wave in a three-dimensional space, respectively. In these cases, the direction of propagation is radial and normal to the surface of the wave. In all these three-dimensional cases, the surfaces that describe the waves are called *wavefronts*. It is usual to determine the properties of the waves by analyzing their wavefronts. This will be evident in the next section, when we study the interference of waves.

It is important to mention that cylindrical and spherical waves can be approached to plane waves. This is due to the expansion of their wavefronts as they propagate. In fact, in many cases, these waves are assumed to be plane waves when the observation distance is very far away from the source, that is, $r \to \infty$.

2.7 Interference of waves

Interference of waves refers to the interaction between two or more waves. In mathematical terms, interference refers to the sum of waves of any nature. As a matter of fact, the general expression of a propagating wave, equation (2.29), can be considered as the interference of an infinite number of waves with spatial frequencies k_n.

2.7.1 Example 1. One-dimensional interference

We can start our analysis by using one-dimensional waves as we did in the previous section. Then, if we assume waves of the same form as in the last section, i.e. $\Psi_n(x,\, t)$, with $n = 1, 2, \ldots$. This means that the interference of N waves can be expressed as:

$$\Psi_{\text{Total}}(x,\, t) = \sum_{n=1}^{N} \Psi_n(x,\, t) \tag{2.38}$$

In order to show a simple example, we can limit the number of waves to $N = 2$, this way we have that $\Psi_{\text{Total}}(x, t) = \Psi_1(x, t) + \Psi_2(x, t)$. Now, if we assume that $\Psi_n(x, t)$ has the same expression as in equation (2.30), then we can see that:

$$\Psi_{\text{Total}}(x, t) = A_{01} \cos(k_1 x - \omega_1 t + \phi_1) + A_{02} \cos(k_2 x - \omega_2 t + \phi_2). \qquad (2.39)$$

Analyzing equation (2.39), it is possible to see that there would be two important cases: the first case is when $k_1 = k_2$ ($\omega_1 = \omega_2$), and the second when $k_1 \neq k_2$ ($\omega_1 \neq \omega_2$). We study these cases in the next sections.

2.7.2 Case 1. $k_1 = k_2$

In this case, we assume that $A_{01} = A_{02} = A_0$, $k_1 = k_2 = k$, $\omega_1 = \omega_2 = \omega$, and $\phi_1 = 0$, then we get:

$$\Psi_{\text{Total}}(x, t) = A_0 \cos(kx - \omega t) + A_0 \cos(kx - \omega t + \phi). \qquad (2.40)$$

If we operate the sum in equation (2.40), we see the following:

$$\Psi_{\text{Total}}(x, t) = 2A_0 \cos\left(\frac{\phi}{2}\right) \cos\left(kx - \omega t - \frac{\phi}{2}\right). \qquad (2.41)$$

Now, in order to observe some interesting findings from equation (2.41), we introduce the amplitude square of the wave $\Psi(x, t)$ as:

$$|\Psi_{\text{Total}}(x, t)|^2 = \Psi_{\text{Total}}(x, t)\Psi_{\text{Total}}(x, t)^*, \qquad (2.42)$$

with $\Psi_{\text{Total}}(x, t)^*$ representing the complex conjugate of $\Psi_{\text{Total}}(x, t)$.

In the particular case we are analyzing here, we assumed $\Psi_{\text{Total}}(x, t) \in \Re$, so its complex conjugate is the same. However, this will become relevant in the next section.

Using equation (2.42), we can see that equation (2.41) has an amplitude square as:

$$|\Psi_{\text{Total}}(x, t)|^2 = 4A_0^2 \cos\left(\frac{\phi}{2}\right)^2 \cos\left(kx - \omega t - \frac{\phi}{2}\right)^2. \qquad (2.43)$$

From equation (2.43), we can see two interesting cases:
- If $\phi = 2m\pi$, with $m \in \mathbb{Z}$, $\Psi(x, t)$ has a maximum value of $4A_0^2 \cos(kx - \omega t)^2$. In fact, this phenomenon is called *constructive interference*.
- If $\phi = (m + \frac{1}{2})\pi$, with $m \in \mathbb{Z}$, then $|\Psi(x, t)|^2 = 0$. Similarly, this is called *destructive interference*.

These cases demonstrate a vital result for waves, that is, waves can interfere in constructive or destructive ways, i.e. there is amplification and reduction of the resulting wave. Figure 2.4 shows the interference between two waves for the extreme cases of ϕ.

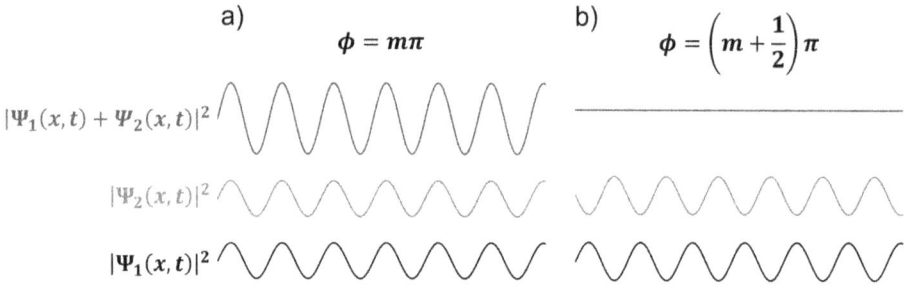

Figure 2.4. Interference between two waves for $\phi = m\pi$ (a) and $\phi = (m + \frac{1}{2})\pi$ (b).

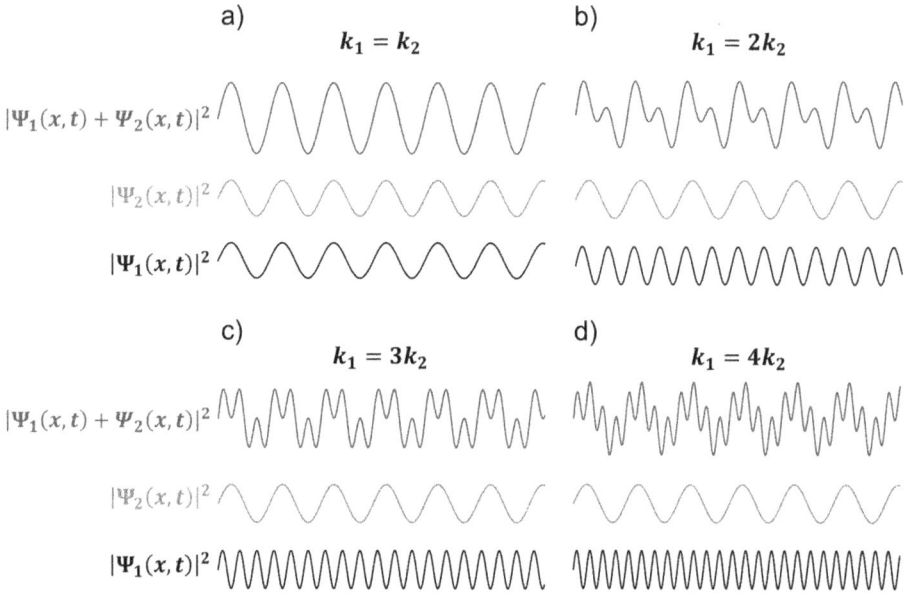

Figure 2.5. Interference between two waves for different relationships between k_1 and k_2, i.e., (a) $k_1 = k_2$, (b) $k_1 = 2k_2$, (c) $k_1 = 3k_2$, and (d) $k_1 = 4k_2$.

2.7.3 Case 2. $k_1 \neq k_2$

As we mentioned above, a different interference effect occurs when the spatial (and temporal) frequencies of $\Psi_1(x, t)$ and $\Psi_2(x, t)$ are different. In such a case, we see that the total wave would have the same expression as equation (2.39). If we now assume that $A_{01} = A_{02} = A_0$, $\phi_1 = \phi_2 = 0$, $t = 0$, then we see that the total wave function looks like this:

$$\Psi_{\text{Total}}(x, t) = A_0(\cos(k_1 x) + \cos(k_2 x)). \tag{2.44}$$

It is clear from equation (2.44) that there will be a modulation in the shape of the wave depending on the difference between k_1 and k_2. Figure 2.5 shows several cases in which k_1 is a factor of k_2.

It is important to note that in the two cases discussed above ($k_1 = k_2$ and $k_1 \neq k_2$), we assumed the time $t = 0$ so we can work only with the spatial frequencies k_i. However, similar results would be obtained if we worked with the temporal frequency ω_i and assumed $x = 0$ instead.

A clear visualization of this case can be viewed in this https://www.youtube.com/shorts/_bV_mEakQ7o video, where acoustic waves were used.

2.7.4 Interference in higher dimensions

Although the cases just analyzed are very instructive, we must pay attention to more general ones. In this sense, when higher dimensions are considered, the phase is less important than the directions of the interacting waves. For example, if we assume a pair of two-dimensional waves defined as $\Psi_1(\mathbf{r}, t)$ and $\Psi_2(\mathbf{r}, t)$, then the interference should be defined as follows:

$$\Psi_{\text{Total}}(\mathbf{r}, t) = \Psi_1(\mathbf{r}, t) + \Psi_2(\mathbf{r}, t). \tag{2.45}$$

If we also assume that these waves are plane waves, the resulting wave is expressed as:

$$\Psi_{\text{Total}}(\mathbf{r}, t) = A_{01}\, e^{-i(\mathbf{k}_1 \cdot \mathbf{r} - \omega t - \phi_1)} + A_{02}\, e^{-i(\mathbf{k}_2 \cdot \mathbf{r} - \omega t - \phi_2)}. \tag{2.46}$$

It is clear from equation (2.47) that the magnitude squared of the wave can be expressed as:

$$|\Psi_{\text{Total}}(\mathbf{r}, t)|^2 = 4A_0^2[1 - \cos^2((\mathbf{k}_1 - \mathbf{k}_2) \cdot \mathbf{r} + \Delta\phi)], \tag{2.47}$$

assuming $A_{01} = A_{02} = A_0$ and $\Delta\phi = \phi_1 - \phi_2$.

Equation (2.47) tells us that the resulting wave will have maxima and minima in the wave magnitude due to the difference of the wave vectors $\mathbf{k}_i$ and the phases ϕ_i. This result is similar to the one we got in section 2.7.1.

Figure (2.6) shows the interference between two plane waves with different wave vectors $\mathbf{k}_i$. Note that the minima and maxima are defined according to the direction of each $\mathbf{k}_i$.

2.8 Electromagnetic waves

We will conclude this chapter by introducing the concept of electromagnetic waves, as this is the primary focus of this book. In this way, electromagnetic fields are solutions to Maxwell equations, the set of fundamental relations that describe the behavior of electromagnetic phenomena. These fields describe how electric and magnetic effects are distributed across space and time and how they influence each other.

If we consider no sources and propagation through vacuum, the Maxwell equations can be written as follows:

$$\nabla \cdot \mathbf{E} = 0, \tag{2.48}$$

$$\nabla \cdot \mathbf{B} = 0, \tag{2.49}$$

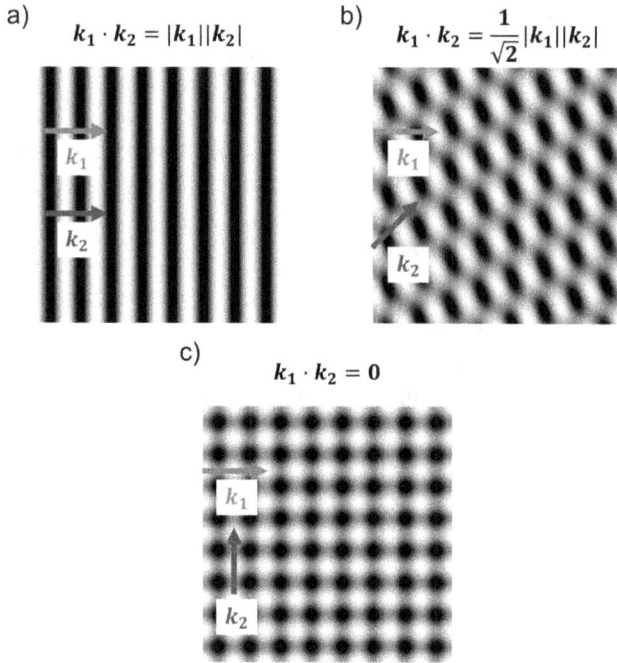

Figure 2.6. Interference between two bi-dimensional plane waves for parallel wave vectors (a), wave vectors at 45° (b), and perpendicular wave vectors (c).

$$\nabla \times \mathbf{E} = -\frac{\partial \mathbf{B}}{\partial t}, \tag{2.50}$$

$$\nabla \times \mathbf{B} = \mu_0 \varepsilon_0 \frac{\partial \mathbf{E}}{\partial t}, \tag{2.51}$$

where equations (2.48) and (2.49) are the Gauss' law for the electric and magnetic fields, respectively, equation (2.50) is Faraday's Law, and equation (2.51) the Ampère–Maxwell law.

It is possible to obtain the wave equation from the Maxwell equations, demonstrating that electromagnetic fields are also waves.

We start by calculating the curl of both sides of Faraday's law, equation(2.50), we get the following:

$$\nabla \times (\nabla \times \mathbf{E}) = -\frac{\partial}{\partial t}(\nabla \times \mathbf{B}). \tag{2.52}$$

Using the fact that $\nabla \times (\nabla \times \mathbf{E}) = \nabla (\nabla \cdot \mathbf{E}) - \nabla^2 \mathbf{E}$ [1], we can see that:

$$\nabla(\nabla \cdot \mathbf{E}) - \nabla^2 \mathbf{E} = -\frac{\partial}{\partial t}(\nabla \times \mathbf{B}). \tag{2.53}$$

We have to remember that we are assuming the electromagnetic field is propagating through a region with no sources, meaning that $\nabla \cdot \mathbf{E} = 0$, according to equation (2.48). Then, the first term to the right of equation (2.53) vanishes.

Finally, replacing $\nabla \times \mathbf{B}$ using equation (2.51), we get:

$$\nabla^2 \mathbf{E} - \mu_0 \varepsilon_0 \frac{\partial^2 \mathbf{E}}{\partial t^2} = 0. \tag{2.54}$$

Similarly, for the magnetic field:

$$\nabla^2 \mathbf{B} - \mu_0 \varepsilon_0 \frac{\partial^2 \mathbf{B}}{\partial t^2} = 0. \tag{2.55}$$

Equations (2.54) and (2.55) confirm that the electric and magnetic fields are waves. Furthermore, these equations show that the speed of an electromagnetic wave (field) in vacuum can be expressed as:

$$c = \frac{1}{\sqrt{\mu_0 \varepsilon_0}} \approx 3.00 \times 10^8 \text{ m s}^{-1}. \tag{2.56}$$

It is possible that the reader recognizes the well-known speed of light c, since light as we know, from gamma rays to the longest radio frequency signals, is an electromagnetic wave.

It is important to note that the solutions to the wave equation shown in section 2.7, equation (2.8), can be also solutions to the wave equations for electromagnetic waves, equations (2.54) and (2.55). This means that it is possible to find electromagnetic plane, cylindrical, and spherical waves. Furthermore, all the treatment for the interference of waves, section 2.7, can be applied to electromagnetic waves. Thus, the wave function $\Psi(\mathbf{r}, t)$ will be replaced now by $\mathbf{E}$ when working with an electric field or $\mathbf{B}$ for a magnetic field[4].

2.8.1 Properties of electromagnetic waves

Electromagnetic waves have some interesting characteristics that will play important roles in the next chapters. For example, an electromagnetic wave propagates in a straight line when the propagating medium is homogeneous and isotropic[5]. This results in the so-called Snell's law [2] when light propagates through an interface between two media of different characteristics.

Another particular characteristic of electromagnetic waves is that the propagation direction is perpendicular to the electric and magnetic fields, it means, $\mathbf{k} \perp \mathbf{E}, \mathbf{B}$. In fact, due to this property, electromagnetic waves are referred as *transverse waves*. Figure 2.7 shows a representation of a propagating electromagnetic wave. The use of a Cartesian reference system helps with the visualization of the wave. A good description of this wave is the following:

---

[4] It is important to remember that $\Psi(\mathbf{r}, t)$ was assumed a scalar function, while $\mathbf{E}$ and $\mathbf{B}$ are vector functions.
[5] This means that there are no preferential directions in the propagating medium.

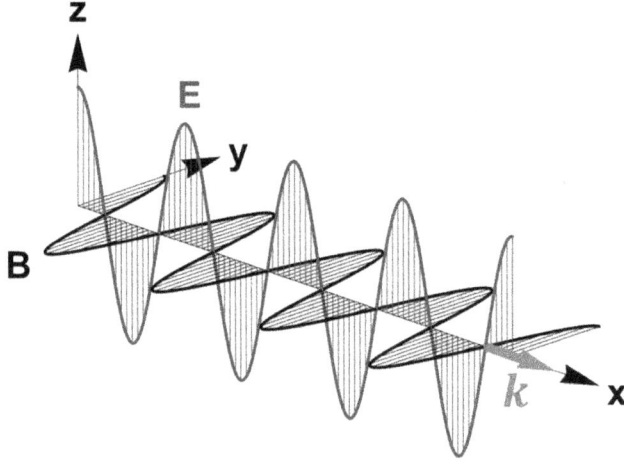

Figure 2.7. Schematic representation of an electromagnetic wave in three dimensions showing **E**, **B**, and **k**. Note that **E** and *H* are not at the correct scale.

$$\mathbf{E}(x, t) = E_0 \cos(kx - \omega t)\hat{z}, \tag{2.57}$$

$$\mathbf{B}(x, t) = B_0 \cos(kx - \omega t)\hat{y}, \tag{2.58}$$

where $\mathbf{k} = k\hat{x}$.

A quite unique property found in electromagnetic waves can be deducted by using the expressions for **E** and **B** in equations (2.57) and (2.58), and substituting them into equation (2.50). It is possible to see that $|\mathbf{E}| = c|\mathbf{B}|$. This means that the light detected in any experiment is mostly the electric field of the wave. As such, we will only focus on **E** throughout the rest of this book.

The transfer of energy produced by an electromagnetic wave occurs in the direction of **k**. Such a direction is also called the *Poynting vector* [2] and is defined as follows:

$$\mathbf{S} = \mathbf{E} \times \mathbf{B}, \tag{2.59}$$

and represents the power delivered by the electromagnetic wave by unit area.

Finally, in section 2.7, we worked with the magnitude square of the wave, $|\Psi(\mathbf{r}, t)|^2$, to show the patterns generated by waves interference. The reason was to simplify the understanding of the so-called *intensity* of an electromagnetic wave, defined as follows:

$$I = \langle \mathbf{S} \rangle = \frac{c\varepsilon_0}{2}|\mathbf{E}(\mathbf{r}, t)|^2, \tag{2.60}$$

with $\langle \mathbf{S} \rangle$ the time average of the Poynting vector and, as mentioned before, $|\mathbf{E}(\mathbf{r}, t)|^2 = \mathbf{E}(\mathbf{r}, t)\mathbf{E}(\mathbf{r}, t)^*$. This means that the intensity is also a way to measure the time average of the power per unit area.

a)   b)

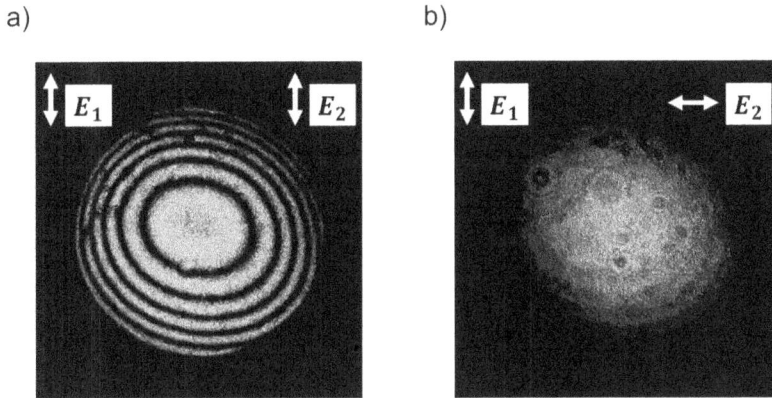

Figure 2.8. Interference pattern between two laser beams with wavelength of 532 nm in a Mach–Zehnder interferometer [2] with the same polarization (a) and perpendicular polarizations (b). Credit: Merlo Lab 2025.

2.8.2 Interference of electromagnetic waves

An important property of electromagnetic waves is their ability to present *polarization*, i.e., a specific direction in which the electric field oscillates, see appendix A. In this sense, electromagnetic waves will interfere only in the case in which two or more waves have the same polarization. Figure 2.8 shows the interference patterns for two cases: when two interacting beams are vertically polarized, figure 2.8(a), and when the beams are polarized in perpendicular direction, figure 2.8(b).

2.8.3 Example 2. Demonstrate that two light beams with perpendicular polarization do not generate interference

We can start by considering two plane waves in phase with the same frequency, ω, and amplitude, E_0, both propagating in the z-direction, and perpendicular polarizations. We can write these fields as:

$$\mathbf{E}_1(z, t) = E_0 \cos(kz - \omega t)\hat{x}, \qquad (2.61)$$

and

$$\mathbf{E}_2(z, t) = E_0 \cos(kz - \omega t)\hat{y}. \qquad (2.62)$$

In order to find the total field (interference between the two beams), we have to add them, then we have:

$$\mathbf{E}_T(z, t) = \mathbf{E}_1(z, t) + \mathbf{E}_2(z, t) = E_0 \cos(kz - \omega t)\hat{x} + E_0 \cos(kz - \omega t)\hat{y}. \qquad (2.63)$$

We finally determine the interference pattern by calculating the intensity in equation (2.63) as $I = \langle |\mathbf{E}_T(z, t)|^2 \rangle$. This way we get:

$$|\mathbf{E}(z, t)|^2 = E_0^2 [\cos^2(kz - \omega t) + \cos^2(kz - \omega t)] + \\ 2E_0^2 \cos(kz - \omega t)\cos(kz - \omega t)\hat{x} \cdot \hat{y}. \qquad (2.64)$$

It is clear from equation (2.64) that the second term vanishes as $\hat{x} \cdot \hat{y} = 0$. In addition, as the intensity is a time average ($\langle \rangle$), the oscillating parts of the first term average to 1, then the intensity is simply $I = 2E_0^2$, meaning there are no fringes in the interference pattern. Figure 2.8 shows the interference pattern between two beams with parallel polarization (a) and perpendicular polarization (b). It is clear that in the case of the perpendicular polarization the intensity is constant.

2.8.4 Example 3. Calculate the interference pattern generated by two point sources

Here, we assume two point sources generating light that can be modeled as spherical waves. If we assume the position of each source as $\mathbf{r}_i$ with $i = 1, 2$, then we can see that the expressions for the fields are:

$$E_1(\mathbf{r}, t) = \frac{A_0}{|\mathbf{r} - \mathbf{r}_1|} e^{i(k|\mathbf{r} - \mathbf{r}_1| - \omega t)}, \tag{2.65}$$

and

$$E_2(\mathbf{r}, t) = \frac{A_0}{|\mathbf{r} - \mathbf{r}_2|} e^{i(k|\mathbf{r} - \mathbf{r}_2| - \omega t)}. \tag{2.66}$$

Note that we have assumed the amplitudes to be A_0. Also, as the field is periodic in time, we can drop the oscillating part $e^{i\omega t}$.

Using equations (2.65) and (2.66), we calculate the total field as:

$$E_T(\mathbf{r}, t) = E_1(\mathbf{r}, t) + E_2(\mathbf{r}, t) = \frac{A_0}{|\mathbf{r} - \mathbf{r}_1|} e^{i(k|\mathbf{r} - \mathbf{r}_1|)} + \frac{A_0}{|\mathbf{r} - \mathbf{r}_2|} e^{i(k|\mathbf{r} - \mathbf{r}_2|)}. \tag{2.67}$$

This means that the intensity will be expressed as:

$$I(\mathbf{r}, t) = |E_1(\mathbf{r}, t) + E_2(\mathbf{r}, t)|^2 = \left| \frac{A_0}{|\mathbf{r} - \mathbf{r}_1|} e^{i(k|\mathbf{r} - \mathbf{r}_1|)} + \frac{A_0}{|\mathbf{r} - \mathbf{r}_2|} e^{i(k|\mathbf{r} - \mathbf{r}_2|)} \right|^2. \tag{2.68}$$

If we call $R_i = |\mathbf{r} - \mathbf{r}_i|$, with $i = 1, 2$, then we equation (2.68) becomes:

$$I(\mathbf{r}, t) = \left| \frac{A_0}{R_1} e^{ikR_1} + \frac{A_0}{R_2} e^{ikR_2} \right|^2. \tag{2.69}$$

As we are assuming the observation of light, we can assume that the observation distance $R \gg \lambda$, meaning that $A_0/R \sim A_0/R_1 \sim A_0/R_2$, then we see that the intensity can be expressed as:

$$I(\mathbf{r}, t) = \left(\frac{A_0}{R} \right)^2 |e^{ikR_1} + e^{ikR_2}|^2. \tag{2.70}$$

Finally, we see that the intensity is expressed as:

$$I(\mathbf{r}, t) = 4 \left(\frac{A_0}{R} \right)^2 \cos^2 \left(\frac{k\Delta R}{2} \right), \tag{2.71}$$

where $\Delta R = R_2 - R_1$. In order to show the physical meaning of equation (2.71), let's assume the following two cases:

- 1. If $\Delta R = m\lambda$, i.e. ΔR is an integer number of wavelengths, we see that the oscillating part of equation (2.71) becomes:

$$\cos^2\left(\frac{k\Delta R}{2}\right) = \cos^2\left(\frac{km\lambda}{2}\right) = \cos^2\left(\frac{2\pi m\lambda}{2\lambda}\right) = \cos^2(m\pi) = 1. \qquad (2.72)$$

This means that the intensity will have a maximum at such a valued of ΔR.

- 2. If $\Delta R = (2m + 1)\lambda/2$, i.e. ΔR is a multiple of one half of a wavelength, we now see that the cos function in equation (2.71) becomes:

$$\cos^2\left(\frac{k\Delta R}{2}\right) = \cos^2\left(\frac{km\lambda}{2}\right) = \cos^2\left(\frac{2\pi\left(\frac{2m+1}{2}\right)\lambda}{2\lambda}\right) = \cos^2\left(\left(\frac{2m+1}{2}\right)\pi\right) = 0. \quad (2.73)$$

This means that the intensity will have a value of zero, or as it is usually called, a minimum.

These results tell us that the intensity will have a periodic set of consecutive maxima and minima, like the one shown in figure 2.6(a). Actually, equation (2.72) matches the famous interference experiment by Thomas Young in 1801, in which it was demonstrated that light is a wave [2].

2.8.5 Practice problem

A Mach–Zehnder interferometer is illuminated by a laser of wavelength λ. A beam splitter separates the beam into two arms of lengths L_1 and L_2. The two beams recombine at the second beam splitter and are projected onto a screen. Derive the expression for the optical path difference (OPD) between the two arms.

2.9 Conclusion

This chapter established the foundation for the understanding of interference of waves. We derived the wave equation, explored its solutions, and showed how boundary conditions lead to different cases, i.e., standing or traveling waves. We introduced the concepts of wavefronts, wave vectors, and phase relationships, which are central to interference phenomena. Finally, we extended these concepts to electromagnetic theory, showing that light follows the same mathematical principles, with some peculiarities as its speed and polarization. All the concepts studied in this chapter set the stage for the next chapter on diffraction, where we analyze the wavefronts behavior from the wave interference standpoint.

References

[1] Arfken G B and Weber H J 1995 *Mathematical Methods for Physicists* (Cambridge, MA: Academic)

[2] Hecht E 2002 *Optics* (Boston, MA: Addison-Wesley)

[3] Khare K, Butola M and Rajora S 2023 *Fourier Optics and Computational Imaging* (Berlin: Springer)

IOP Publishing

Optical Interference and Dynamic Diffraction
Research methods for undergraduates
Jenny Magnes and Juan M Merlo-Ramírez

Chapter 3

Diffraction

3.1 Introduction

This chapter introduces the theoretical framework required to describe the diffraction of waves. We begin by presenting the Huygens–Fresnel principle, which interprets every point on a wavefront as a secondary source of spherical wavelets. Building on the general wave equation developed previously, we derive mathematical models for diffraction using the scalar formulation.

Next, we develop analytical approaches such as Kirchhoff's integral and the Fresnel–Kirchhoff diffraction formula and applied to canonical problems, including circular apertures and wires.

We conclude by discussing near-field (Fresnel) and far-field (Fraunhofer) regimes, as well as the Fourier transform perspective that underpins modern diffraction analysis that is the core of this book.

3.2 Diffraction basics

Diffraction is a fundamental wave phenomenon that describes the deviation of waves from their initial propagation direction when they encounter obstacles or apertures comparable in size to their wavelength [1]. For example, figure 3.1(a) shows the diffraction pattern generated by a beam of electrons passing through a thin layer of graphite and figure 3.1(b) shows the diffraction of a water wave interacting with a grating.

The last means that we could describe a diffraction theory based on the general case of waves using $\Psi(\mathbf{r}, t)$, as we did with interference theory in section 2.7. However, we will focus only in the case of diffraction of light, i.e. electromagnetic fields. This is because this book is precisely devoted to the study diffraction of light.

doi:10.1088/978-0-7503-4836-2ch3 3-1

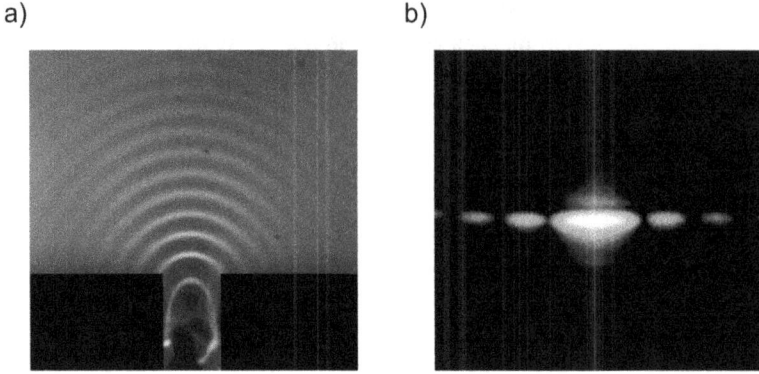

Figure 3.1. (a) Diffraction of water waves by a slit. (b) Diffraction pattern generated by a slit when illuminated with a laser diode with wavelength of 532 nm. Credit: Merlo Lab 2025.

3.3 Huygens–Fresnel principle

Christiaan Huygens proposed in 1690 that any wavefront of light is the result of the superposition of a series of point sources [1], called *wavelets*. This idea was revolutionary as it suggested that light is a wave, contradicting Newton's idea of light corpuscles [2].

In this sense, we could use the mathematical formulation described in chapter 2 to describe the superposition of waves, as:

$$\Psi_{\text{Total}}(\mathbf{r},\, t) = \sum_{i=1}^{N} \Psi_i(\mathbf{r},\, t), \tag{3.1}$$

with $\mathbf{r}$ the position of each point source and $\Psi(\mathbf{r},\, t)$ a spherical wave.

3.3.1 Example 1. Plane wavefront generated by a set of point sources

In order to illustrate this interesting case, we can use a two-dimensional system to better visualization. In this case, we assume that each bidimensional *spherical* wave has the following expression[1]:

$$\Psi_i(\mathbf{r}_i,\, t) = \frac{\Psi_0}{r_i} e^{-i(kr_i \,-\, \omega t)}, \tag{3.2}$$

where Ψ_0 is the amplitude of the wave.

If we assume that the wave is periodic in time, then we could simplify equation (3.2) as:

$$\Psi_i(\mathbf{r},\, t) = \frac{\Psi_0}{r_i} e^{-i(kr_i)}. \tag{3.3}$$

---

[1] It is clear that in order to display this waves in a two-dimensional space, we need to assume the observation is done at the plane $z = 0$.

a) **3 Sources**

b) **5 Sources**

c) **21 Sources**

d) **400 Sources**

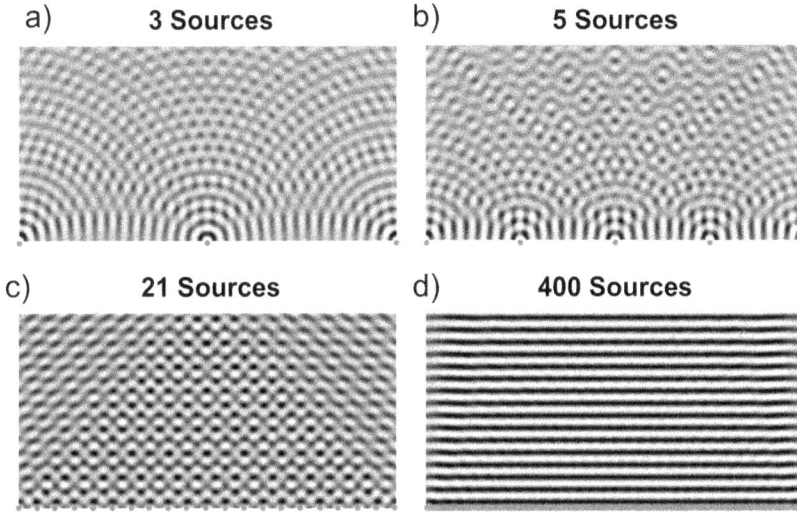

Figure 3.2. Superposition of point sources for different numbers of sources. (a) Three sources, (b) five sources, (c) twenty-one sources, and (d) four hundred sources.

Now, it is very important to understand that equation (3.3) is useful only when we know the exact position of each source $\mathbf{r}$. So, for this example, we can assume that $\mathbf{r} = x\hat{x}$, with $-\infty \leqslant x \leqslant \infty$, and the total wave looks like:

$$\Psi_i(\mathbf{r},\, t) = \sum_{i=1}^{N} \frac{\Psi_0}{r_i} e^{-i(kr_i)}. \tag{3.4}$$

Figure 3.2 shows the superposition of several sets of two-dimensional point sources. Although it could be obvious, it is interesting to note that as the number of sources increases, the wavefront resembles more the actual proposed wavefront.

3.3.2 Example 2. Converging wavefront generated by a set of point sources

In this case, we can still use equation (3.4); however, we need to impose a condition to the position of the point sources, i.e., the sources must be located at the edge of a circular sector. In this sense, the position of the sources could be expressed as $\mathbf{r} = r_0 \cos \theta \hat{x} + r_0 \sin \theta \hat{y}$, with r_0 the radius of the array, and θ the aperture angle. The resulting wavefront is shown in figure 3.3; a different number of point sources are indicated to demonstrate the effect on the total wavefront.

In figure 3.3, it is interesting to see how the increase of point sources resulted in the familiar wave behavior we could expect in a converging case. Moreover, note how there is a minor change among the cases of 20 and 100 sources. This is due to the geometry of the sources array.

It is clear from examples in sections 3.3.1 and 3.3.2 that we can generalize equation (3.1) as:

$$\Psi_{\text{Total}}(\mathbf{r}) = \sum_{i=1}^{\infty} \frac{\Psi_0}{r_i} e^{-i(kr_i)}. \tag{3.5}$$

a) **5 Sources**

b) **10 Sources**

c) **20 Sources**

d) **100 Sources**

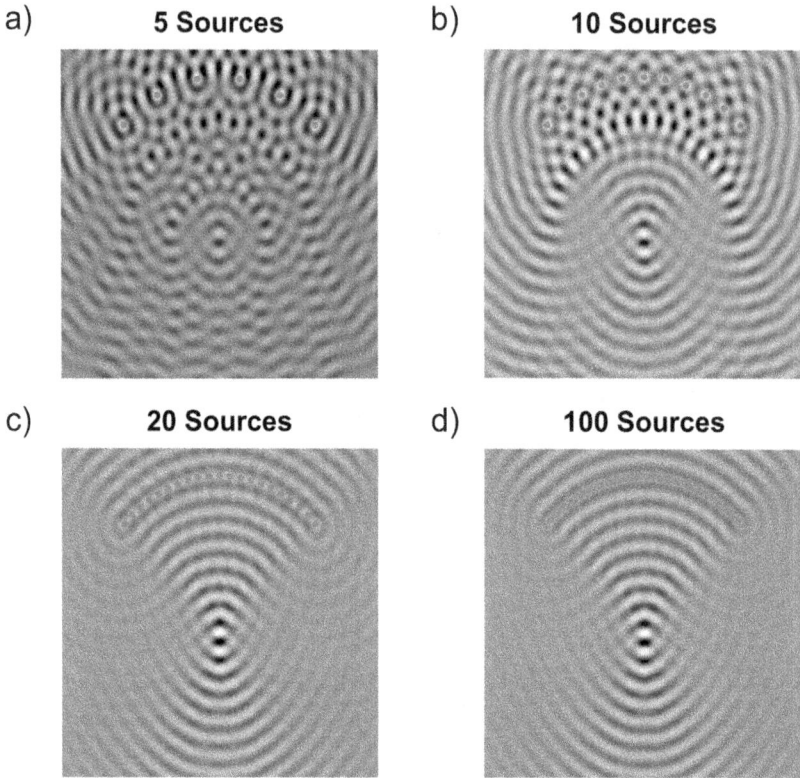

Figure 3.3. Superposition of point sources for a curved wavefront for $\pi/4 \leqslant \theta \leqslant 3\pi/4$. (a) Five sources, (b) ten sources, (c) twenty sources, and (d) one hundred sources.

Equation (3.5) means that the distance between the sources must approach zero to correctly reproduce a wavefront[2]. However, even when this description of wavefronts is simple and easy to implement, we will be always limited by the *discreteness* of the sources[3]. Thus, we would need a more general approach to the matter. In this sense, the next section is dedicated to discussing a mathematical method that describes better the physics of diffraction.

3.4 Diffraction theory

In the chapter 2, we defined a wave as perturbation evolving in space and time, $\Psi(\mathbf{r}, t)$. We could express this wave in a general way as:

---

[2] In practical cases, this condition could be achieved by using a distance that is considered much smaller than the wavelength. For example, in figure 3.2, the distance between sources was set to 0.1λ, with λ the wavelength. Note that although the wavefront approaches well a flat one in figure 3.2(d), there are still remains of the point source distribution.

[3] In the next chapter, we describe the use of the continuum version of equation (3.5).

$$\Psi(\mathbf{r},\, t) = \Psi_0(\mathbf{r})e^{-i(\omega t)}, \tag{3.6}$$

with $\Psi_0(\mathbf{r})$ the spatial part of the wave.

Now, inserting equation (3.6) into the wave equation, (2.8), we could see that we get:

$$e^{-i\omega t}\left(\nabla^2\Psi(\mathbf{r}) + k^2\Psi(\mathbf{r})\right) = 0, \tag{3.7}$$

where $k = 2\pi/\lambda$ is the wave number.

Finally, we see that equation (3.7) can be written as:

$$\nabla^2\Psi(\mathbf{r}) + k^2\Psi(\mathbf{r}) = 0, \tag{3.8}$$

also known as the *Helmholtz equation*.

There are several methods to find the solution to the Helmholtz equation, (3.8). The most relevant for the topic we are treating here is the Kirchhoff formulation [3].

Due to the scope of this book, we can assume that the analysis of equation (3.11) is done over an circular aperture. Although this would mean the we must analyze a three-dimensional system, we could focus only in a plane. Figure 3.4 shows the schematic representation of our single aperture analysis.

The Kirchhoff formulation to calculate the wave at the point P, see figure 3.4, can be written as follows:

$$\Psi(P) = \int_S \left(G\frac{\partial\Psi}{\partial n} - \Psi\frac{\partial G}{\partial n}\right) dS, \tag{3.9}$$

where G is a Green's function, i.e., the generator of waves.

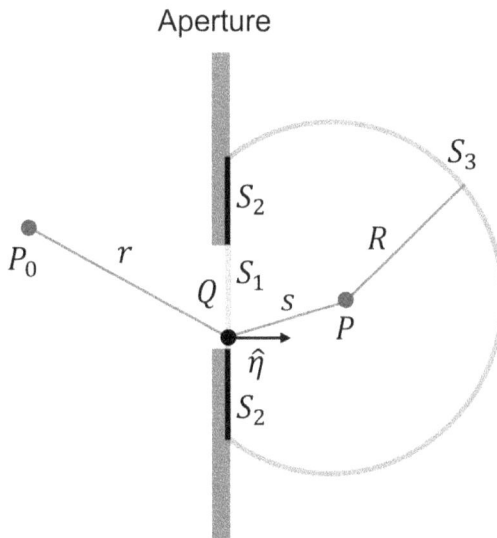

Figure 3.4. Scheme for the analysis of the Kirchhoff theory of diffraction.

For the particular case we are treating here, we could set the Green's function to be a point source defined as:

$$G(s) = \frac{e^{iks}}{s}. \tag{3.10}$$

In the same way, we could assume that $\Psi(\mathbf{r})$ has the same form as equation (3.3). Figure 3.4 shows the meaning of s and r.

Substituting the $G(s)$ and $\Psi(\mathbf{r})$ into equation (3.9), we get:

$$\Psi(\mathbf{P}) = \int_S \left(\frac{e^{iks}}{s} \frac{\partial \Psi}{\partial n} - \Psi \frac{\partial}{\partial n}\left(\frac{e^{iks}}{s}\right) \right) \, \mathrm{d}S, \tag{3.11}$$

Using the scheme shown in figure 3.4, we can see that the integral over the surface S in equation (3.11) can be split into three separate integrals, i.e., as follows:

$$\Psi(P) = \left(\int_{S_1} + \int_{S_2} + \int_{S_3} \right) \left(\frac{e^{iks}}{s} \frac{\partial}{\partial n}\left(\frac{e^{ikr}}{r}\right) - \left(\frac{e^{ikr}}{r}\right) \frac{\partial}{\partial n}\left(\frac{e^{iks}}{s}\right) \right) \, \mathrm{d}S. \tag{3.12}$$

Now, in order to solve equation (3.12), we can use the following assumptions:

1. The wave, Ψ, at the aperture S_1 is the same as the one without the aperture, i.e., $\Psi = \Psi_0 \frac{e^{ikr}}{r}$. This means:

$$\frac{\partial \Psi}{\partial n} = \Psi_0 \frac{e^{ikr}}{r}\left(ik - \frac{1}{r}\right) \cos \chi, \tag{3.13}$$

   with χ the angle between r and $\hat{\eta}$.
2. The wave is null at the surface S_2, meaning $\Psi = 0$ and $\partial \Psi/\partial n = 0$.[4]
3. We could assume that R, see figure 3.4, is large, so the integral over the surface S_3 does not contribute to the total wave.

Next, we calculate the normal derivative on e^{iks}/s at the point Q, see figure 3.4, and we get:

$$\frac{\partial}{\partial n}\left(\frac{e^{iks}}{s}\right) = \left(\frac{e^{iks}}{s}\right)\left(ik - \frac{1}{s}\right)\cos \theta, \tag{3.14}$$

where θ is the angle between s and $\hat{\eta}$.

Then, we could see that equation (3.12) can be written as:

$$\Psi(P) = \frac{\Psi_0}{4\pi} \int_{S_1} \frac{e^{ik(s+r)}}{rs}\left[\cos \chi \left(ik - \frac{1}{r}\right) - \cos \theta \left(ik - \frac{1}{s}\right) \right] \mathrm{d}S \tag{3.15}$$

---

[4] This assumption suggests that the wave is discontinuous at the edges of the aperture. However, for our purposes here, we could use it.

Equation (3.15) is the general case for the diffraction of a wave generated by an aperture of any shape. However, in order to locate our findings on the research of diffraction of light, we can assume that optical systems have dimensions much larger than the wavelength of light. This means that $\lambda \ll r, s$, thus $(ik - \frac{1}{r}) \rightarrow ik$ and $(ik - \frac{1}{s}) \rightarrow ik$. Then, we can finally write (3.15) as:

$$\Psi(P) = \frac{i\Psi_0}{2\lambda} \int_{\text{Aperture}} \frac{e^{ik(s+r)}}{rs}(\cos \chi - \cos \theta)\mathrm{d}S \tag{3.16}$$

Equation (3.16) is the so-called *Fresnel–Kirchhoff diffraction formula* and it is a simple way to determine the diffraction pattern generated by an aperture at any distance from the aperture's plane.

An important point to be mentioned is that equation (3.16) is a generalization of equation (3.5).

3.4.1 Example 3. Determine the diffraction pattern generated by a circular aperture

In order to facilitate the analysis of this example, we provide figure 3.5, where an aperture is showed. The wave source is called P_0 and the observation plane is located at P. A point, where secondary wavelets are generated, is located in the aperture and is called Q. The auxiliary distances are called r' and s' and these distances do not depend of the position of Q.

As we mentioned before, the cases analyzed in this book are framed in the context of diffraction of light in an optical system, in which all the distances and dimensions of optical components are much larger than the wavelength of light. As such, we can assume that $\frac{1}{2}(\cos \chi - \cos \theta) \rightarrow \cos \delta$, with δ the angle formed between the line joining P_0 and P with the normal of the aperture's plane. In addition, we can

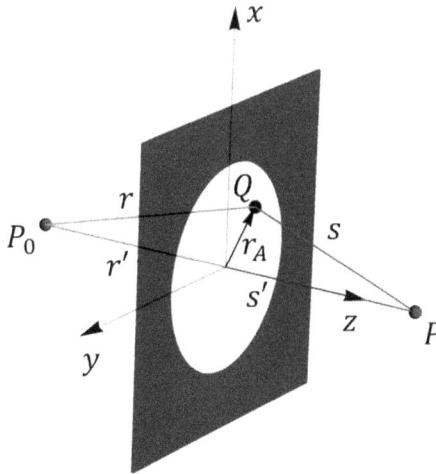

Figure 3.5. Scheme for the analysis of the diffraction generated by a circular aperture.

consider that the diameter of the aperture is much smaller than the distance to the light source and the plane of observation, i.e., $a \ll r, s$. This condition is known as the *far-field*.

Under our assumptions, we can write equation (3.16) in the following way:

$$\Psi(P) = \frac{i\Psi_0}{\lambda} \frac{\cos \delta}{r's'} \int_{\text{Aperture}} e^{ik(s+r)} dS. \tag{3.17}$$

We have used r' and s' here due to the far-field condition. It is clear from figure 3.5 that r' and s' do not depend on the position of Q.

In order to simplify our analysis, we could set a coordinate system at the center of the aperture, see figure 3.5. Under these conditions, we could use the the auxiliary r' and s' to relate the position of the point Q, r_A, in the surface of the aperture, with the position of source P_0 and the observation plane P. If we call P_0: (x_0, y_0, z_0), P: (x, y, z), and r_A: (X_A, Y_A), we can see that:

$$r^2 = (x_0 - X_A)^2 + (y_0 - Y_A)^2 + z_0^2, \tag{3.18}$$

and

$$s^2 = (x - X_A)^2 + (y - Y_A)^2 + z^2. \tag{3.19}$$

In figure 3.5, it is clear that if $r'^2 = x_0^2 + y_0^2 + z_0^2$ and $s'^2 = x^2 + y^2 + z^2$, then we can express equations (3.18) and (3.19) as follow:

$$r^2 = r'^2 - 2(x_0 X_A + y_0 Y_A) + X_A^2 + Y_A^2, \tag{3.20}$$

and

$$s^2 = s'^2 - 2(x X_A + y Y_A) + X_A^2 + Y_A^2. \tag{3.21}$$

We mentioned before that $a \ll r, s$. In such a case, we could expand equations (3.20) and (3.21) in terms of X_A/r', Y_A/r', X_A/s', and Y_A/s' to get:

$$r \sim r' - \frac{x_0 X_A + y_0 Y_A}{r'} + \frac{X_A^2 + Y_A^2}{2r'} - \frac{(x_0 X_A + y_0 Y_A)^2}{2r'^3} + \cdots, \tag{3.22}$$

and

$$s \sim s' - \frac{x X + y Y}{s'} + \frac{X^2 + Y^2}{2s'} - \frac{(x X + y Y)^2}{2s'^3} + \cdots, \tag{3.23}$$

If we keep only the linear terms, we see that $r \sim r' - \frac{x_0 X_A + y_0 Y_A}{r'}$ and $s \sim s' - \frac{x X_A + y Y_A}{s'}$. Using these approximations, equation (3.17) can be written as follows:

$$\Psi(P) = \frac{i\Psi_0}{\lambda} \frac{\cos \delta}{r's'} e^{ik(r'+s')} \int_{\text{Aperture}} e^{-ik\left(\frac{x_0 X_A + y_0 Y_A}{r} + \frac{x X_A + y Y_A}{s}\right)} dS. \tag{3.24}$$

If we rearrange the argument of the exponential function in the integral as:

$$\frac{x_0 X_A + y_0 Y_A}{r'} + \frac{x X_A + y Y_A}{s'} = \left(\frac{x_0}{r'} + \frac{x}{s'}\right) X_A + \left(\frac{y_0}{r'} + \frac{y}{s'}\right) Y_A, \qquad (3.25)$$

then, equation (3.24) looks as:

$$\Psi(P) = \frac{i\Psi_0}{\lambda} \frac{\cos\delta}{r's'} e^{ik(r'+s')} \int_{\text{Aperture}} e^{-ik\left(\left(\frac{x_0}{r'} + \frac{x}{s'}\right)X_A + \left(\frac{y_0}{r'} + \frac{y}{s'}\right)Y_A\right)} dS. \qquad (3.26)$$

If we call $f_x = \frac{1}{\lambda}(\frac{x_0}{r'} + \frac{x}{s'})$ and $f_y = \frac{1}{\lambda}(\frac{y_0}{r'} + \frac{y}{s'})$, then equation (3.26) becomes[5]:

$$\Psi(P) = \frac{i\Psi_0}{\lambda} \frac{\cos\delta}{r's'} e^{ik(r'+s')} \int_{\text{Aperture}} e^{-2\pi i(f_x X_A + f_y Y_A)} dS. \qquad (3.27)$$

Equation (3.27) represents the diffracted wave by an aperture and measured at point P. Note how the aperture has arbitrary shape as we have not introduced any restriction. In this sense, we can make a change of coordinates to facilitate the solution of the problem. Due to the symmetry of the problem, the best choice would be to use polar coordinates as follows:

$$X_A = \rho\cos\phi, \qquad Y_A = \rho\sin\phi, \qquad (3.28)$$

with $\rho \in [0, a]$ and $\phi \in [0, 2\pi]$.

If we also define $f_x = \rho'\cos\alpha$ and $f_y = \rho'\sin\alpha$, then the argument in the exponential of equation (3.27) becomes:

$$f_x \cos\phi + f_y \sin\phi = \rho\rho'\cos(\phi - \alpha). \qquad (3.29)$$

Introducing equation (3.29) into equation (3.27), we get:

$$\Psi(P) = \frac{i\Psi_0}{\lambda} \frac{\cos\delta}{r's'} e^{ik(r'+s')} \int_0^a \int_0^{2\pi} e^{-2\pi i\rho\rho'\cos(\phi-\alpha)} \rho\, d\rho\, d\phi. \qquad (3.30)$$

Now, we can see from equation (3.30) that the term outside the integral is a constant that would not change if we alter the size of the aperture. Then, we could simply approach it as a constant Ψ_A. With this in mind, we see that equation (3.30) can be written as:

$$\Psi(P) = \Psi_A \int_0^a \rho\, d\rho \int_0^{2\pi} e^{-2\pi i\rho\rho'\cos(\phi-\alpha)} d\phi. \qquad (3.31)$$

In equation (3.31), we see that the angle α is an arbitrary phase that does not depend on the shape or the size of the aperture, then we could ignore it. Then, the second integral of equation (3.31) would have a familiar solution, as follows:

$$\int_0^{2\pi} e^{-2\pi i\rho\rho'\cos(\phi)} d\phi = 2\pi J_0(\rho'\rho), \qquad (3.32)$$

---

[5] It is possible that an experienced reader could relate the integral in equation (3.27) to a Fourier transform. Actually, in the following chapters we will use this methodology for the calculation of diffraction patterns.

with J_0 being the zeroth-order Bessel function of the first kind. This means that equation (3.31) is now written as:

$$\Psi(P) = \Psi_A \int_0^a \rho \, d\rho (2\pi J_0(\rho'\rho)).$$ (3.33)

We know that the Bessel function has the following property:

$$\int_0^a \rho J_0(K\rho) \, d\rho = \frac{aJ_1(Ka)}{K},$$ (3.34)

then we can see that equation (3.33) becomes:

$$\Psi(P) = \Psi_A \frac{aJ_1(2\pi a\rho')}{2\pi\rho'}.$$ (3.35)

We have defined $\rho' = \sqrt{f_x^2 + f_y^2}$, meaning that $\rho' = \rho'(r', s')$. In order to facilitate the solution of equation (3.35), we could assume that the source, r', is located at infinity, meaning that $\rho' \rightarrow \frac{1}{\lambda}\sqrt{(\frac{x}{s'})^2 + (\frac{y}{s'})^2}$. In addition, if we assume that the distance to the observation point, P', is much larger than the size of the diffraction pattern, then we could see that $r' \rightarrow z$, then $\rho' \rightarrow \frac{1}{\lambda}\sqrt{(\frac{x}{z})^2 + (\frac{y}{z})^2} = \frac{1}{\lambda z}\sqrt{x^2 + y^2}$. Then, we can see that $R = \sqrt{x^2 + y^2}$ is the radial distance in the plane where P is located. This means that equation (3.35) can be written as:

$$\Psi(P) = \Psi_A \frac{J_1(2\pi a\frac{R}{\lambda z})}{2\pi\frac{R}{\lambda z}}.$$ (3.36)

Up to this point, we have carried out our calculations on the amplitude of the diffracted wave $\Psi(P)$, however this is a complex function. Thus, we should calculate the intensity of this wave to find the *diffraction pattern* generated by the circular aperture. Then, if we multiply equation (3.36) by its complex conjugate, we finally get:

$$I(P) = 2I_A \left(\frac{J_1(2\pi a\frac{R}{\lambda z})}{2\pi\frac{R}{\lambda z}} \right)^2.$$ (3.37)

Figure 3.6 shows the diffraction pattern generated by a circular aperture using equation (3.37), figure 3.6(a), and an experimental setup, figure 3.6(b). In both cases, the wavelength was 632 nm. It is interesting to see how accurate is equation (3.37) reproducing the experimental observation. This is why the methodology described in this section has been used and applied for the last two hundred years.

3.4.2 Practice problem

Calculate the diffraction pattern generated by a square aperture. Assume the light source is located at infinity and the dimensions of the diffraction pattern are much smaller than the distance from the observation point to the aperture's plane.

3.5 Diffraction of electromagnetic fields

The last section has been dedicated to the study of diffraction of a wave by an aperture of any shape. However, it is possible that the reader is wondering why we used a scalar theory, if light is an electromagnetic field, i.e., light needs a vector to be described, see section 2.8. In this sense, there exists the so-called *vectorial diffraction theory*, in which light is described by its components, it means, $\mathbf{E}(\mathbf{r}, t)$ and $\mathbf{H}(\mathbf{r}, t)$. Despite this, we focus only in the scalar theory for two reasons: first, the scope of this book is on the study of diffraction patterns studied only with the intensity of light as described by equation (3.37); second, in experimental optics, it is common practice to use laser light, see section A.4, that is usually linearly polarized. This means that $\mathbf{E} = E_0 \hat{x}_i$, with $\hat{x}_i = \hat{x}, \hat{y}, \hat{z}$, thus, the field has only one component and consequently the application of the scalar theory is valid[6].

With this in mind, we can rewrite equation (3.9) as:

$$E(P) = \int_S \left(G \frac{\partial E}{\partial n} - E \frac{\partial G}{\partial n} \right) dS. \tag{3.38}$$

An important point in equation (3.38) is that the function G should not be necessarily a point source, as described in equation (3.10). It can be any kind of wave.

a) **Calculated**   b) **Experimental**

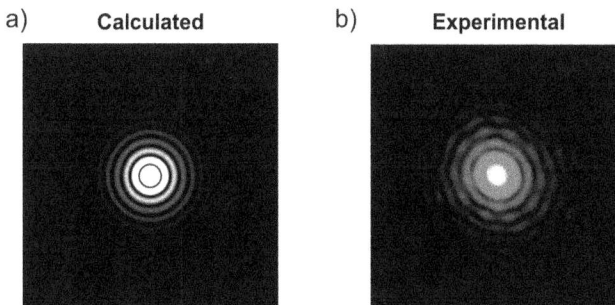

Figure 3.6. (a) Calculated and (b) experimental diffraction pattern generated by a circular aperture (red) laser light from a helium neon (HeNe) laser. The wavelength in both cases was 632 nm. Credit in (b): Merlo Lab 2025.

---

[6] If the reader is interested in the rigorous diffraction theory, it is advised to see chapter 9 of reference [4].

3.5.1 Example 4. Determine the diffraction pattern generated by a thin wire illuminated by a plane wave

In order to solve this problem, we will use *Babinet's principle*, that states that the diffracted field by an aperture, $E_A(P)$, is related to the diffracted field by the complementary shape to the aperture, $E_{A'}(P)$, and the field without obstruction, E_0, as follows:

$$E_\circ(P) = E_A(P) + E_{A'}(P). \tag{3.39}$$

For this particular case, we can assume that field without obstruction is a plane wave $E_\circ = E_0 e^{ikz}$, assuming the field propagates in the z-direction. Now, we need to calculate the diffracted field by the complementary of a wire, that would be a rectangular slit with infinite dimensions in the vertical direction. Then, let's assume the slit has a width $2a$ in the x-direction and extends infinitely in the y-direction.

We could use the equation (3.38) to solve this problem for the slit with $E_\circ = E_0 e^{ikr}/r$ and $G = e^{ikr}/r$; we get this:

$$E_{\text{Slit}}(P) = \frac{iE_{0S}}{\lambda} \frac{e^{iks'}}{s'} \int_{\text{Aperture}} e^{-2\pi i(f_x X_A + f_y Y_A)} dX_A dY_A. \tag{3.40}$$

As in this case, the aperture is infinite in the y-direction, we can drop the integral in dy, to get:

$$E_{\text{Slit}}(P) = \frac{iE_{0S}}{\lambda} \frac{e^{iks'}}{s'} \int_{-a}^{a} e^{-2\pi i f_x X_A} dX_A. \tag{3.41}$$

In this case, $f_x = \frac{1}{\lambda}(\frac{x}{z})$, then equation (3.41) becomes:

$$E_{\text{Slit}}(P) = \frac{iE_{0S}}{\lambda} \frac{e^{iks'}}{s'} \int_{-a}^{a} e^{-ik\frac{x}{z} X_A} dX_A. \tag{3.42}$$

After the evaluation of equation (3.42), the diffracted field by the slit is simply:

$$E_{\text{Slit}}(P) = \frac{iE_{0S}}{\pi} e^{ikz} \text{sinc}\left(\frac{kax}{z}\right). \tag{3.43}$$

Applying this finding to equation (3.39), we get:

$$E_{\text{Wire}}(P) = E_\circ(P) - E_{\text{Slit}}(P) \Rightarrow E_{\text{Wire}} = E_0 e^{ikz} - \frac{iE_{0S}}{\pi} e^{ikz} \text{sinc}\left(\frac{kax}{z}\right). \tag{3.44}$$

If we assume the fields have the same amplitude, i.e., $E_0 = E_{0S}$, we can see that the field of the diffraction pattern generated by the wire takes this form:

$$E_{\text{Wire}} = E_0 e^{ikz}\left(1 - \frac{i}{\pi}\text{sinc}\left(\frac{kax}{z}\right)\right). \tag{3.45}$$

Finally, the intensity of the diffraction pattern is:

$$I_{\text{Wire}} = I_0\left(1 - \frac{1}{\pi}\text{sinc}\left(\frac{kax}{z}\right)\right)^2. \tag{3.46}$$

It is clear from example in section 3.4.1 that Babinet's principle is a powerful tool to determine the diffracted field by any structure, always such a structure is modeled as the complementary of an aperture.

3.6 Fresnel diffraction

Up to this point, we have mentioned that the diffraction theory used in the rest of this book is based on the Fraunhofer diffraction (far-field). However, in order to complete our discussion about diffraction, we need to talk about the Fresnel approximation, also known as *near-field approximation.*

In order to show the difference between the near- and far-field approximations, the next example is dedicated to calculate the diffraction pattern generated by a circular aperture under the Fresnel approximation. For a complete discussion of the Fresnel theory of diffraction, the reader is advised to check chapter 9 of reference [4].

3.6.1 Example 4. Determine the diffraction pattern generated by a circular aperture under the Fresnel approximation

When discussing the far-field approximation, we kept only the linear terms in equation (3.22) and equation (3.23), that resulted in equation (3.27). Contrarily, when we use the near-field approach, we can keep the quadratic terms in equation (3.22) and equation (3.23); under these circumstances, equation (3.27) takes the following form:

$$E(P) = E_A \int_0^a \rho e^{\frac{ik\rho^2}{z}}[2\pi J_0(\rho'\rho)]d\rho, \tag{3.47}$$

where E_A is defined as:

$$E_A = \frac{2\pi E_0}{i\lambda z}e^{ikz}e^{\frac{ik\rho'^2}{2z}}. \tag{3.48}$$

It is clear that equation (3.47) would reduce to equation (3.33) if we assume $\rho' \ll z$. So, we can say that the Fraunhofer approximation is a special case of the Fresnel approximation.

On the other hand, equation (3.47) does not have an analytical solution. Thus, a numerical solution is needed to observe the diffraction pattern at any point in the z-direction. Figure 3.7 shows the cross-section of the diffraction of light generated by a circular aperture, calculated by the finite element method [5][7]. It is interesting to note the differences in the diffracted field between the Fresnel, figure 3.7, and the

---

[7] The finite element method uses a rigorous vectorial analysis of the diffraction. This means that the numerical results obtained by this method have no assumptions as the ones made in this chapter.

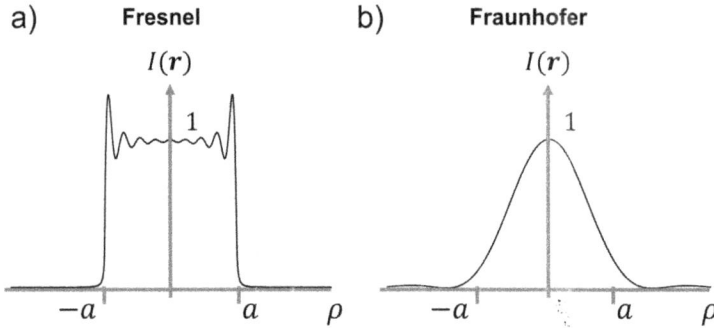

Figure 3.7. Cross-section of the diffracted field generated by a circular aperture in the Fresnel (a) and Fraunhofer (b) approximation.

Fraunhofer, figure 3.6(a), approximations. It is clear that the near-field, Fresnel, region has a richer set of interactions. This is due to the existence of higher order terms in the exponential function of equation (3.47).

3.7 Diffraction pattern as a transform

Up to this point, we have described the diffraction pattern using equation (3.9) under some assumptions. In this sense, we obtained equation (3.27) for the case in which the distances r' and s' do not depend on the aperture's shape or characteristics. Furthermore, equation (3.27) is dependent on f_x and f_y, whose notation was chosen to represent spatial frequencies related to the wave vector $\mathbf{k}$.

Now, it would be interesting to rewrite equation (3.27), but this time assuming that we do not know the original shape of the source of secondary wavelets, $E(x, y)$, and performing all the procedure for the Fraunhofer approximation, we would get the following:

$$E(P) \propto \int A(x, y)E(x, y)e^{-2\pi i(f_x X_A + f_y Y_A)}dS, \qquad (3.49)$$

with $A(x, y)$ a function containing the information of the aperture.

It is important to mention that in equation (3.49), the product $A(x, y)E(x, y)$ is equivalent to what we called the Green's function in section 3.4. Also, the form of (3.49) is considered a transform; in this particular case, in which we used the spatial *frequencies* to describe it, it is called the *Fourier transform* [6].

Finally, as the function $A(x, y)$ contains all the information related to the aperture, we need to explore all the frequencies available in the space, it means, we need to integrate f_x and f_y from $-\infty$ to ∞. Then, we can write the formal definition of the diffraction pattern as follows:

$$E(P) = E_0 \int_{-\infty}^{\infty} \int_{-\infty}^{\infty} A(x, y)E(x, y)e^{-2\pi i(f_x X_A + f_y Y_A)}dX_A dY_A. \qquad (3.50)$$

Equation (3.50) has profound implications. Particularly for the calculation of diffraction patterns as most of the apertures used in optics experiments can be modeled as simple functions. Let's put to work equation (3.50) to show this.

3.7.1 Example 5. Calculate the diffraction pattern generated by a slit. Assume the slit is illuminated by a plane wave

A single slit can be modeled analytically with a rectangular function with explicit form $A(x, y) = \text{rect}(x/a)$, with a the width of the slit. As the function depends only on x, we can drop the y-dependency. Also, if we assume the illumination is a plane wave, we simply define $E(x, y) = 1$, then the diffraction pattern can be calculated as:

$$E(P) = E_0 \int_{-\infty}^{\infty} \text{rect}\left(\frac{x}{a}\right) e^{-2\pi i f_x X_A} \, dX_A = \mathscr{F}\left[\text{rect}\left(\frac{x}{a}\right)\right], \tag{3.51}$$

where $\mathscr{F}$ is the symbol for the Fourier Transform.

It can be found any that $\mathscr{F}[\text{rect}(\frac{x}{a})] = \text{sinc}(\frac{kax}{z})$ [6], then we finally get:

$$E(P) \propto \text{sinc}\left(\frac{kax}{z}\right), \tag{3.52}$$

that is the same expression as equation (3.43), up to a constant.

Finally, we can express the diffraction pattern as intensity as follows:

$$I(P) \propto \text{sinc}^2\left(\frac{kax}{z}\right). \tag{3.53}$$

It is clear that equation (3.52) has the same form as equation (3.43). This means that the Fourier transform method results in the same answer as using the Fresnel–Kirchhoff theory, discussed in section 3.4.

3.7.2 Example 6. Calculate the diffraction pattern generated by a pair of circular apertures with the same radius, R, and separated by a distance d. Use the Fraunhofer approximation

In the Fraunhofer approximation, we know that the diffraction pattern is proportional to the Fourier transform of the aperture function.

We can assume that the each aperture is modeled with the following function[8]:

$$\Pi\left(\frac{r}{R}\right) = \begin{cases} 1, & \text{if } r \leqslant R, \\ 0, & \text{if } r > R. \end{cases} \tag{3.54}$$

If we suppose the position of each aperture is defined as $\mathbf{r}_1 = -\frac{d}{2}\hat{x}$ and $\mathbf{r}_2 = \frac{d}{2}\hat{x}$, then it is clear that the total aperture function can be expressed as:

---

[8] This function represents an open circle located at $r = 0$ and radius R.

$$t(\mathbf{r}) = \Pi\left(\frac{|\mathbf{r} - \mathbf{r_1}|}{R}\right) + \Pi\left(\frac{|\mathbf{r} - \mathbf{r_2}|}{R}\right). \tag{3.55}$$

This means that the diffraction pattern will be expressed as:

$$E(\mathbf{r}) = \mathscr{F}[t(\mathbf{r})], \tag{3.56}$$

or in other words:

$$E(\mathbf{k}) = \mathscr{F}\left[\Pi\left(\frac{|\mathbf{r} - \mathbf{r_1}|}{R}\right) + \Pi\left(\frac{|\mathbf{r} - \mathbf{r_2}|}{R}\right)\right]. \tag{3.57}$$

At this point, we can use the *shift theorem* of the Fourier Transform[9][7] to get:

$$E(\mathbf{k}) = E_{single}(\mathbf{k})(e^{-i\mathbf{k}\cdot\mathbf{r_1}} + e^{-i\mathbf{k}\cdot\mathbf{r_2}}), \tag{3.58}$$

where $E_{single}(\mathbf{k})$ is the diffraction pattern generated by a single circular aperture and $\mathbf{k} \cdot \mathbf{r}_{1,2} = \pm\frac{\pi d}{\lambda}\sin\theta$. It is easy to see that:

$$E(\theta) = 2E_{single}(\theta)\cos\left(\frac{\pi d}{\lambda}\sin\theta\right). \tag{3.59}$$

We saw in the example in section 3.4.1 that the diffraction pattern generated by a single circular aperture is:

$$E_{single}(\theta) \propto 2E_A(\theta)\left(\frac{J_1(2\pi a\frac{R}{\lambda z})}{2\pi\frac{R}{\lambda z}}\right). \tag{3.60}$$

Due to the Fraunhofer approximation, equation (3.60) is equivalent to:

$$E_{single}(\theta) \propto 2E_A(\theta)\left(\frac{J_1(kR\sin\theta)}{kR\sin\theta}\right). \tag{3.61}$$

If we substitute equation (3.61) into equation (3.59), we get:

$$E(\theta) \propto 2E_0(\theta)\left(\frac{J_1(kR\sin\theta)}{kR\sin\theta}\right)\cos(2kd\sin\theta). \tag{3.62}$$

Finally, the intensity of the diffraction pattern is expressed as follows:

$$I(\theta) \propto 2I_0(\theta)\left(\frac{J_1(kR\sin\theta)}{kR\sin\theta}\right)^2\cos^2(2kd\sin\theta), \tag{3.63}$$

where $k = \frac{2\pi}{\lambda}$. Figure 3.8 shows the diffraction pattern generated by two circular apertures with a diameter of 10 μm separated by a distance of 50 μm.

---

[9] This will be discussed in chapter 6.

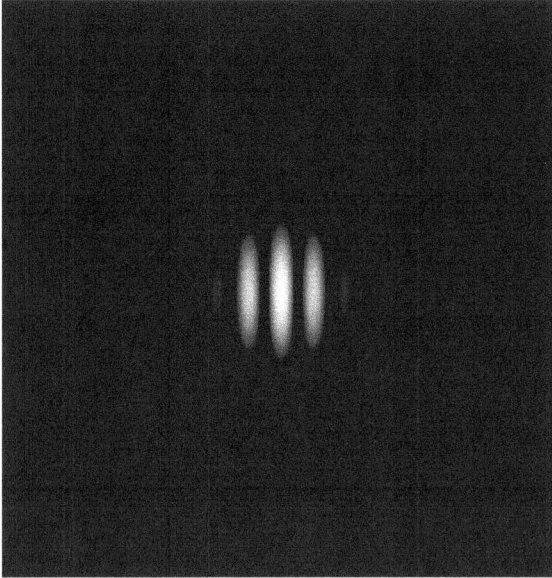

Figure 3.8. Diffraction pattern generated by two circular apertures with diameter of 10 μm separated by a distance of 50 μm. Both apertures were illuminated by light with 660 nm wavelength.

3.8 Conclusion

In this chapter, we developed a comprehensive description of diffraction, showing how wavefronts evolve when encountering apertures and obstacles. We derived analytical expressions for diffraction patterns and distinguished between near-field and far-field regimes. Finally, we demonstrated how the diffraction of light can be interpreted as a Fourier transform of the aperture function, providing a powerful tool for analyzing complex optical systems, as we will discuss in upcoming chapters.

References

[1] Hecht E 2002 *Optics* (Boston, MA: Addison-Wesley)
[2] Aspect A 2017 From Huygens' waves to Einstein's photons: Weird light *C. R. Physique* **18** 498–503
[3] Chandra S and Sharma M K 2024 *A Textbook of Optics* (Berlin: Springer)
[4] Born M and Wolf E 1999 *Principles of Optics: Electromagnetic Theory of Propagation, Interference and Diffraction of Light* 7th edn (Cambridge: Cambridge University Press)
[5] Sheng X Q and Song W 2012 *Essentials of Computational Electromagnetics* (New York: Wiley)
[6] Khare K, Butola M and Rajora S 2023 *Fourier Optics and Computational Imaging* (Berlin: Springer)
[7] Goodman J W 2005 *Introduction to Fourier optics* (Greenwood Village, CO: Roberts and Company Publishers)

IOP Publishing

Optical Interference and Dynamic Diffraction
Research methods for undergraduates
Jenny Magnes and Juan M Merlo-Ramírez

Chapter 4

Fraunhofer (far-field) diffraction

4.1 Introduction

How does an object interact with an electromagnetic wave? When a light wave is incident on a surface, it can be scattered, absorbed, or reflected. Here, we explore the scattering of an electromagnetic wave. In particular, we will focus on coherent monochromatic light and consider its distribution in the far-field. Monochromatic light is limited to one wavelength, and coherent light assumes all light is in phase. Of course, experimental reality dictates that there is some uncertainty in all properties, such as wavelength and phase.

Far-field diffraction, also known as Fraunhofer diffraction, is based on several assumptions: (a) the dimensions of the diffracting object are much smaller[1] than the distance from the object to the screen where light is observed; this is also called the far-field approximation. (b) The diffracting object must be comparable in size to the wavelength used for interference effects to become apparent. (c) Only diffraction at small angles is consistent with our calculations, known as the small-angle approximation.

Fourier transforms can be used to calculate diffraction patterns analytically. We will demonstrate how the patterns are closely related to Fourier transforms and use this to develop the effects of an object's translation and rotation upon its far-field diffraction pattern. Then we can compare the calculations with the actual data.

This chapter will introduce the analytical expression for calculating an object's diffraction pattern. However, in practice, diffraction patterns are mainly modeled computationally since analytical methods get very complicated. In the next chapter, we will introduce computational methods.

---

[1] 'Much smaller' or 'much larger' usually refer to at least a factor of ten (one order of magnitude) difference.

doi:10.1088/978-0-7503-4836-2ch4

4.2 Young's double slit

In 1801, Thomas Young demonstrated the wave properties of light using an experimental setup with a card separating a sunbeam [1] into two beams that formed a far-field diffraction pattern. This experiment was later adapted to what is commonly known as Young's double-slit experiment [2] and is typically presented in introductory physics texts [3, 4].

We briefly describe Young's double-slit experiment and outline the calculations to illustrate the underlying assumptions in chapters 2 and 3. These descriptions will also facilitate the introduction of dynamic diffraction. In this chapter, we highlight that the same assumptions used to calculate the location of constructive and destructive interference in Young's double slit are valid for Fraunhofer diffraction in general. In other words, Young's double slit is a special case of Fraunhofer diffraction.

4.3 Fraunhofer diffraction

Consider a diffracting object with features comparable in size to the scattering wavelength λ as illustrated in an example in figure 4.1. After the scattering occurs, the diffraction pattern is observed—typically on a screen—at a distance D from the object. This is distance d much larger than the size of the scattering object.

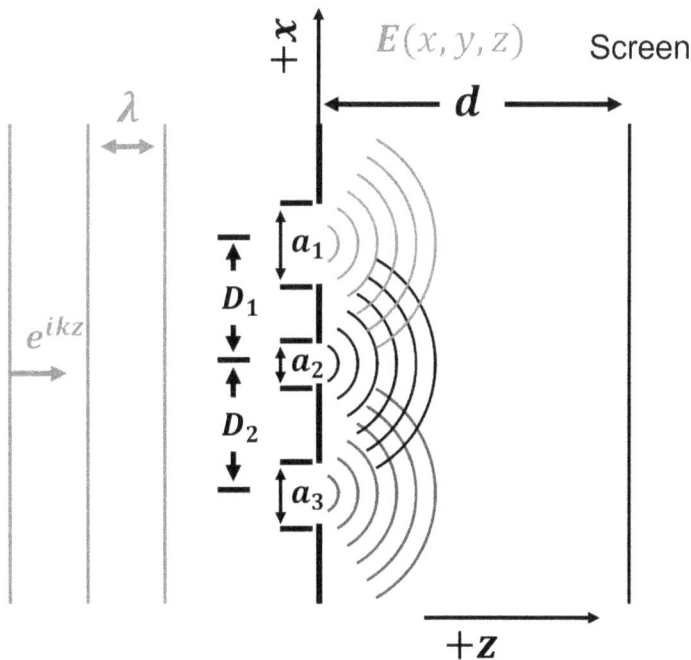

Figure 4.1. An incoming plane wave with wavelength λ on the left diffracts on an object with slit widths a_1, a_2, and a_3 that are separated by distances D_1 and D_2 respectively. The observation points are represented by the screen at a distance d from the diffracting object on the right.

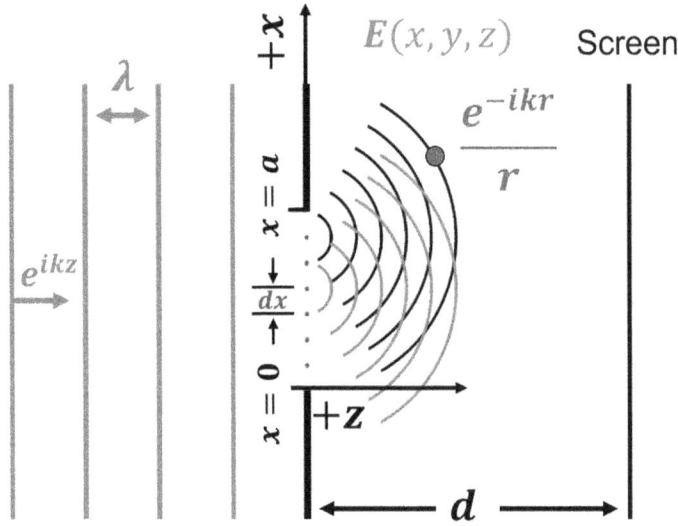

Figure 4.2. The single slit itself can be broken down into sections of dx, where dx is an infinitesimally small section serving as a wave source.

Figure 4.2 illustrates a single slit and its application of Huygen's principle; every point is the source of a wave. The wave on the right is a spherical wave with an amplitude that decreases by $\frac{e^{-ikr}}{r}$ as the wave propagates a distance r from its origin with a wave vector $k = 2\pi/\lambda$. Each infinitesimal section dx in the slit serves as a source for a wave.

We apply the principle of superposition to every point on the screen. We sum up all the electric field amplitudes by integrating over the width of the slit so that the electric field:

$$E = E_0 \int_{x=0}^{x=a} \frac{e^{-i\mathbf{k}\cdot\mathbf{r}}}{r} dx, \qquad (4.1)$$

where E_0 is the electric field amplitude, x is the distance from the edge of the slit, and r is the distance from a point in the slit opening from which a wavelet originates to the screen (figure 4.2).

Let x_s be the distance from $x = 0$ on the screen to the point on the screen where we want to identify the strength of the electric field (figure 4.3); we can then express r as the distance to any point on the screen in terms of x:

$$r(x) = \sqrt{(x - x_s)^2 + d^2}, \qquad (4.2)$$

where d is the separation between the slit and the screen. We factor out d to make some approximations that simplify the integral when we substitute equation (4.2) in equation (4.1):

$$r(x) = d\sqrt{\left(\frac{x - x_s}{d}\right)^2 + 1}, \qquad (4.3)$$

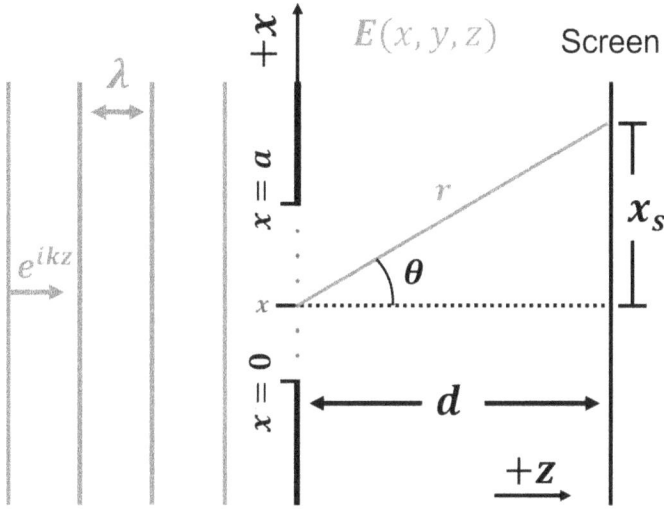

Figure 4.3. The distances are not to scale. The slit width a is much smaller than the distance from the origin to the point at which we are calculating the diffraction pattern.

We can now apply the small-angle approximation to the case when $x - x_s \ll d$ so that the first term under the square root in equation (4.3) becomes a small number ε:

$$r(\varepsilon) = d\sqrt{(\varepsilon + 1)}, \tag{4.4}$$

We then expand equation (4.4) using the Taylor expansion [5] around $\varepsilon = 0$:

$$r(\varepsilon) = r(0) + \frac{r'(0)}{1!}\varepsilon + \frac{r''(0)}{2!}\varepsilon^2 + \dots \tag{4.5}$$

We only keep the first two terms, since the subsequent terms are much smaller than the first two terms Dropping these terms turns Taylor's expansion into an approximation:

$$r(\varepsilon) \approx d + \frac{1}{2}\mathrm{d}\varepsilon = d + \frac{1}{2}\frac{(x - x_s)^2}{d}. \tag{4.6}$$

With this approximation, equation (4.1) can be modified by acknowledging how small the term $\frac{(x - x_s)^2}{d}$ is compared to d. We can assume that the denominator is approximately d. We cannot make the same assumption for the exponent, since it grows fast and k is a large value:

$$E = E_0 \int_{x=0}^{x=a} e^{-ikd} \frac{e^{-ik_x \frac{1}{2}\frac{(x-x_s)^2}{d}}}{d + \frac{1}{2}\frac{(x - x_s)^2}{d}} dx \approx E_0 \frac{1}{d} e^{-ikd} \int_{x=0}^{x=a} e^{-ik_x \frac{1}{2}\frac{(x-x_s)^2}{d}} dx. \tag{4.7}$$

The last term in the approximation is called the Fresnel integral. Now we will expand $(x - x_s)^2$ into $x^2 + x_s^2 - 2xx_s$. Recall that $x \ll x_s$, since x will always be between 0 and a. For optical diffraction, the slit width is typically the width of a human hair or less, while the pattern on the screen is large enough to be easily visible to the human eye, i.e., a few inches. Applying this approximation:

$$x^2 + x_s^2 - 2xx_s \approx x_s^2 - 2xx_s. \tag{4.8}$$

In addition, we can use the small-angle approximation $\tan \theta \approx \sin \theta$ (figure 4.3) for angles much less than 1 radian so that:

$$k\frac{x_s}{d} = k \tan \theta \approx k \sin \theta = k_x. \tag{4.9}$$

Our integral now becomes:

$$E = E_0 \frac{1}{d} e^{-ikd} e^{-ik\frac{x_s^2}{2d}} \int_{x=0}^{x=a} e^{ik_x x} dx. \tag{4.10}$$

We can modify this integral by inserting any shape described by an aperture function $g(x)$ between 0 and a. It does not have to be a single slit. Also, the terms before the integral are constants and can be absorbed in the amplitude. For those reasons, we can write:

$$E = E_0 \int_{x=0}^{x=a} g(x) e^{ik_x x} dx \tag{4.11}$$

The aperture function could extend from a negative to a positive value. Of course, the limits only have to extend where the aperture function is not equal to zero. For the general case, the Fourier integral[2] becomes:

$$E = E_0 \int_{x=-\infty}^{x=+\infty} g(x) e^{ik_x x} dx \tag{4.12}$$

The equation above is the Fourier transform of the aperture function, which gives us a general way of calculating the diffraction pattern. Our eyes and other optical sensors do not detect the electric field directly. Optical sensors detect intensity I, which is proportional to the electric field magnitude squared. In general, the magnitude squared of the Fourier transform is called the 'power spectrum' or 'spectral power density' [6]:

$$I \propto |\mathbf{E}|^2 = \mathbf{E}^*\mathbf{E}, \tag{4.13}$$

where $\mathbf{E}^*$ is the complex conjugate of the electric field $\mathbf{E}$. Looking back at equation (4.10) the magnitude squared of e^{-ikd} turns out to be one since:

$$|e^{-ikd}|^2 = e^{ikd} e^{-ikd} = e^0 = 1. \tag{4.14}$$

---

[2] In other fields such as signal processing or electrical engineering, the convention may dictate that the Fourier integral carries a negative sign in the exponent $E = E_0 \int_{x=0}^{x=a} g(x) e^{-ik_x x} dx$. In that case, the inverse Fourier transform will carry a positive sign.

similarly $e^{-ik\frac{x_s^2}{2d}}$ equals 1 in magnitude. In other words, while the phase factors matter during the interference process, when electric fields are superimposed, they do not matter when calculating the final intensity distribution.

4.3.1 Example: single slit

A common example for calculating a far-field diffraction pattern is a single slit of width a. Let's imagine a slit of width a. The electric field E has a uniform amplitude of E_0 across the slit. We can model this situation using a top-hat function of width a as the aperture function $g(x)$:

$$g(x) = \Pi_a = \begin{cases} 0, & -\infty < x < -a/2 \\ 1, & -a/2 < x < a/2 \\ 0, & a/2 < x < \infty \end{cases} \tag{4.15}$$

Calculating the electric field at a point on the screen:

$$E = E_0 \int_{x=-a/2}^{x=a/2} e^{ik_x x} dx = E_0 \frac{e^{ik_x x}}{ik_x} \Big|_{-a/2}^{a/2} = E_0 \frac{e^{ik_x a/2} - e^{-ik_x a/2}}{ik_x} \tag{4.16}$$

The integration limits match the slit from $-a/2$ to $a/2$ since the aperture function is equal to zero elsewhere. Let's use Euler's formula:

$$e^{iy} = \cos y + i \sin y \Rightarrow \sin y = \frac{e^{iy} - e^{-iy}}{2i}, \tag{4.17}$$

where y is a number. Now,

$$E = E_0 2i \frac{\sin(k_x a/2)}{ik_x} = E_0 a \,\mathrm{sinc}\,(k_x a/2). \tag{4.18}$$

There is no position parameter such as x instead the variable becomes k_x. We say that the Fourier transform is represented in k-space while the aperture function—in this case a single slit—exists in the object space. The two spaces are reciprocal to each other. A larger a; i.e., a large slit, causes a larger frequency in the sinc function resulting in a shorter wavelength—a more narrow sinc function. Of course, you might argue that the pattern exists on the screen in a space defined by distances rather than a wave number k_x. We can reintroduce our approximation from equation (4.9) containing the distance on the screen x_s

$$E = E_0 a \,\mathrm{sinc}\,(k\frac{x_s}{d}a/2). \tag{4.19}$$

For the last step, we need to calculate the intensity profile I, which is proportional to $|E|^2$:

$$I \propto a^2 \,\mathrm{sinc}^2\left(\frac{k x_s a}{2d}\right). \tag{4.20}$$

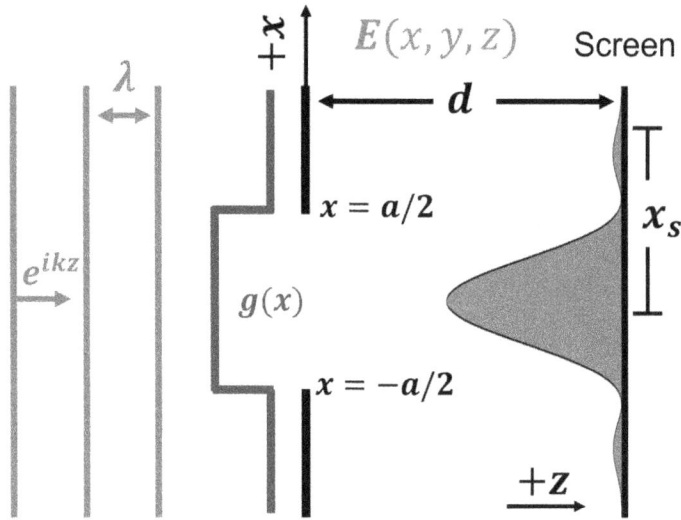

Figure 4.4. A single slit represented by the aperture function $g(x)$. The diffraction pattern is illustrated on the right where the peaks pointing to the left represent constructive interference and the valleys represent dimmer areas; i.?e., destructive interference. The distances are not to scale. The slit width a is much smaller than the distance from the origin to the point at which we are calculating the diffraction pattern.

This is the intensity profile illustrated in figure 4.4. It is the familiar diffraction pattern of a single slit.

4.3.2 Practice problem

Young's double slit: two slits are spaced a distance b from center to center. Each slit has a width of a. Calculate the electric field strength E in the far-field diffraction pattern in terms of E_0, k_x, a, and b. Plot the intensity versus $k_x/2\pi$. Is your result consistent with the previous results showing maxima when $m\lambda \approx b \sin \theta$, where m refers to the mth order maximum?

4.4 Conclusion

We have explored how to calculate a far-field or Fraunhofer diffraction pattern. In that context, we focus on the approximations that lead to a reasonable far-field diffraction pattern given an arbitrary aperture function. We saw that a symmetric aperture function leads to a real-valued Fourier transform. In the next chapter, we will explore more how an asymmetric aperture function leads to an imaginary-valued FT. More complicated aperture functions might lead to more difficult solutions, or no analytical solution might exist. In the next chapter, we will explore how we can calculate the diffraction pattern due to any aperture function using computational means.

References

[1] Young T 1804 I. The Bakerian Lecture. Experiments and calculations relative to physical optics *Phil. Trans. R. Soc.* **94** 1–16

[2] Martienssen W and Spiller E 1964 Coherence and fluctuations in light beams *Am. J. Phys.* **32** 919–26

[3] Halliday D, Resnick R and Walker J 2013 *Fundamentals of Physics* (New York: Wiley)

[4] Knight R D 2017 *Physics for Scientists and Engineers: A Strategic Approach with Modern Physics* (Boston, MA: Pearson)

[5] Arfken G B and Weber H J 1995 *Mathematical Methods for Physicists* (Cambridge, MA: Academic)

[6] James J F 2011 *A Studentas Guide to Fourier Transforms: With Applications in Physics and Engineering* (Cambridge: Cambridge University Press)

IOP Publishing

Optical Interference and Dynamic Diffraction

Research methods for undergraduates

Jenny Magnes and Juan M Merlo-Ramírez

Chapter 5

Computing diffraction patterns

5.1 Introduction

In the previous chapter, we learned how to calculate a far-field diffraction pattern analytically due to an arbitrary aperture function; i.e., the shape of the diffracting object. Two significant challenges must be overcome for an analytical calculation: (1) an analytical expression of the aperture function must be available; (2) an analytical solution must exist. Sometimes, an analytical solution may exist, but it can be tedious and time-consuming. The calculation of complicated far-field diffraction patterns is now possible due to advances in computing power using discrete Fourier transforms (DFTs). A computer is not capable of analytic calculations, but it can repeat many discrete calculations very quickly.

Interestingly, the DFT as we know it was developed by Johann Carl Friedrich Gauss during the early 19th century [1] (published after Gauss's death) before the analytical Fourier transform (FT) was developed by Jean-Baptiste Joseph Fourier [2].

In this chapter, we will demonstrate how a DFT functions in principle. This involves matrix methods and numerous summation signs. We will use visual aids wherever possible to deepen the understanding of FTs in general. We will begin with one-dimensional (1D) DFTs and then expand to the two-dimensional (2D) case. We will illustrate computational considerations as the number of terms for our calculations increases.

5.2 One-dimensional discrete Fourier transform

First, we need a discrete aperture function (object) to compute the diffraction pattern of an object. That means that we have to discretize the object. For the 1D case, we can just imagine an array of values that represents the aperture function. We can write these values in an $N \times 1$ matrix, where N is the number of matrix elements:

doi:10.1088/978-0-7503-4836-2ch5

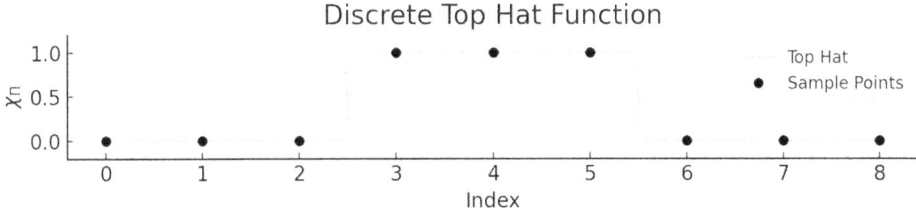

Figure 5.1. A representation of the top hat function χ_Π using a matrix. The yellow line indicates the discrete approximation by the points in black.

$$\chi = (\chi_0\ \chi_1\ \chi_2\ \cdots\ \chi_N), \tag{5.1}$$

As a familiar example, we can represent the top hat function in this way (figure 5.1):

$$\chi_\Pi = (0\ 0\ 0\ 1\ 1\ 1\ 0\ 0\ 0), \tag{5.2}$$

Now, we need an expression that helps us transform the discrete points. A 1D DFT is similar to the analytic version, except that the integral sign is replaced by a summation sign. In this way, each transformed matrix element X_m reads:

$$X_m = \frac{1}{N}\sum_{n=0}^{N-1}\chi_n e^{2\pi i n m/N}, \tag{5.3}$$

where m is the mth matrix element in the DFT X and n refers to the nth matrix element in the aperture function χ.

Inserting $m = 0$ and $N = 9$ for our specific example χ_Π, the first matrix element X_0 becomes:

$$\begin{aligned}
X_0 &= \frac{1}{9}\sum_{n=0}^{9-1}\chi_n e^{2\pi i n0/9} = \frac{1}{9}\sum_{n=0}^{9-1}\chi_n e^0 = \frac{1}{9}\sum_{n=0}^{9-1}\chi_n 1 = \frac{1}{9}\sum_{n=0}^{9-1}\chi_n \\
&= \frac{1}{9}(0 + 0 + 0 + 1 + 1 + 1 + 0 + 0 + 0) = 3/9 = 1/3.
\end{aligned} \tag{5.4}$$

For $m = 1$, X_1 becomes:

$$\begin{aligned}
X_1 &= \frac{1}{9}\sum_{n=0}^{9-1}\chi_n e^{2\pi i n1/9} \\
&= \frac{1}{9}(0 \cdot e^{2\pi i0\cdot 1/9} + 0 \cdot e^{2\pi i1\cdot 1/9} + 0 \cdot e^{2\pi i2\cdot 1/9} + 1 \cdot e^{2\pi i3\cdot 1/9} \\
&\quad + 1 \cdot e^{2\pi i4\cdot 1/9} + 1 \cdot e^{2\pi i5\cdot 1/9} + 0 \cdot e^{2\pi i6\cdot 1/9} + 0 \cdot e^{2\pi i7\cdot 1/9} + 0 \cdot e^{2\pi i8\cdot 1/9}) \\
&= \frac{1}{9}(0 + 0 + 0 + 1 \cdot e^{2\pi i3\cdot 1/9} + 1 \cdot e^{2\pi i4\cdot 1/9} + 1 \cdot e^{2\pi i5\cdot 1/9} + 0 + 0 + 0) \\
&= \frac{1}{9}(e^{2\pi i/3} + e^{2\pi i4/9} + e^{2\pi i5/9}).
\end{aligned} \tag{5.5}$$

Table 5.1. Discrete Fourier transform values X_m of χ_{Π} with squared magnitudes.

m	$\mathrm{Re}(X_m)$	$\mathrm{Im}(X_m)$	$|X_m|^2$
0	‾0.3333	‾0.0000	0.1111
1	−0.2644	‾0.0962	0.0781
2	‾0.1147	−0.0962	0.0225
3	−0.0000	‾0.0000	0.0000
4	−0.0170	‾0.0962	0.0095
5	−0.0170	−0.0962	0.0095
6	‾0.0000	‾0.0000	0.0000
7	‾0.1147	‾0.0962	0.0225
8	−0.2644	−0.0962	0.0781

Using Euler's identity, equation (4.17):

$$X_1 = \frac{1}{9}\left(\cos\frac{2\pi}{3} + i\sin\frac{2\pi}{3} + \cos\frac{2\pi \cdot 4}{9} + i\sin\frac{2\pi \cdot 4}{9} + \cos\frac{2\pi \cdot 5}{9} + i\sin\frac{2\pi \cdot 5}{9}\right) \quad (5.6)$$
$$= -0.2644 + 0.0962i.$$

Similarly, the other matrix elements can be calculated as listed in table 5.1. We encourage the reader to verify these values. Some values repeat in the table. We will expand on this phenomenon later.

Plotting $|X_m|^2$, we expect to see something that resembles a sinc function. However, as can be seen in figure 5.2(a), we would need to cut the graph in half, between the matrix elements $m = 4$ and $m = 5$, and then tape the ends together to obtain something that peaks in the middle, resembling a sinc function. We can also modify equation (5.3) and shift m by $\frac{N-1}{2}$:

$$X_m = \frac{1}{N}\sum_{n=-\frac{N-1}{2}}^{\frac{N-1}{2}} \chi_n e^{2\pi i n m/N}, \quad \text{for odd N}$$

$$X_m = \frac{1}{N}\sum_{n=-\frac{N}{2}}^{\frac{N}{2}} \chi_n e^{2\pi i n m/N}, \quad \text{for even N.}$$

$$(5.7)$$

To calculate diffraction patterns, equation (5.7) is applicable since we are probing the aperture function (object) from positive to negative infinity, where infinity refers to the span of the object. Equation (5.3) is used for the frequency analysis of time-dependent communication signals, such as radio signals, wifi signals, or signals transmitted over fiber optics.

Using the expression for odd N in equation (5.7), the shape vaguely resembles a sinc function in figure 5.2(b). The shape looks choppy, or, in other words, unresolved. In the next section, we will explore the concept of resolution in DFTs.

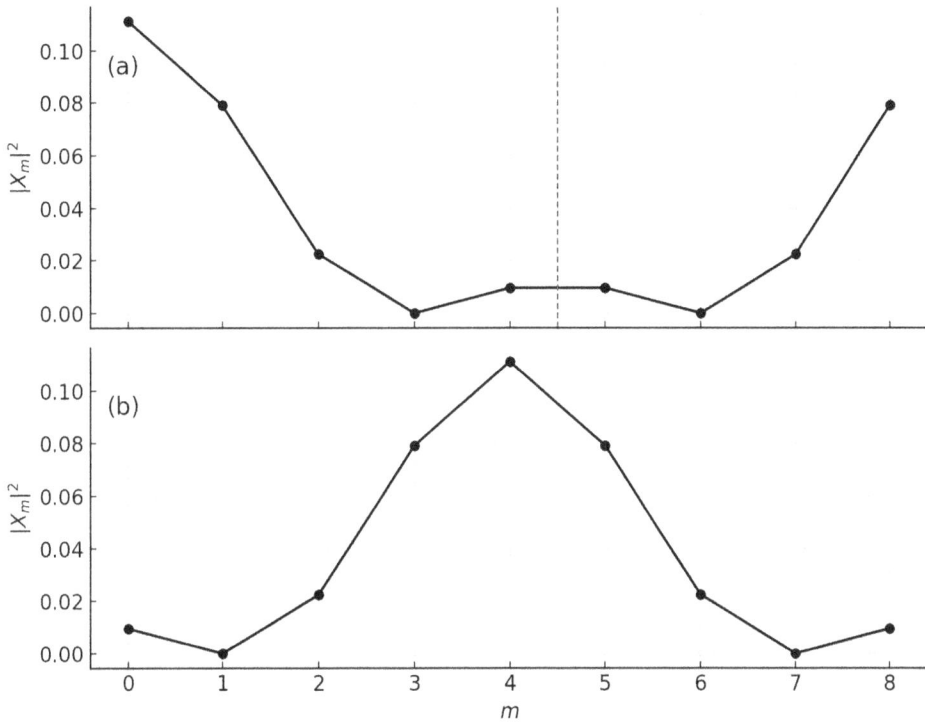

Figure 5.2. The magnitude squared of the DFT of the top hat function χ; i.e., X (a) using equation (5.3), and (b) using equation (5.7) with m shifted by $\frac{N-1}{2}$ representing an outline of the diffraction pattern.

5.3 Resolution in discrete Fourier transforms

When we discretize an aperture function, we have to make some choices about how many points to include to resolve the object. We also have to decide how many zero points to include on either side of the aperture. For example, if we omitted zeros on each side of the top hat in the top hat function, the top hat would not be recognizable. Let us start by doubling the number of zeros on each side of the top hat function, so that instead of three zeros on each side, we have six zeros on each side (figure 5.3(a)). We have now increased the density of points for the entire DFT, and the central maximum and first minimum are beginning to emerge clearly. This phenomenon is known as oversampling.

Had we not included the points around the aperture, we would not have known that we have an aperture. Not only that, we must sample enough points around the aperture to resolve the DFT. How many points do we need? The Nyquist frequency dictates that the number of oversampled points must be more than half of the total number of points [3].

You might ask how this increased resolution comes about. Remember from equation (5.6) that our exponential terms $e^{2\pi i n(m-(\frac{N-1}{2})/N)}$ stand for a combination of

Top Hat Function

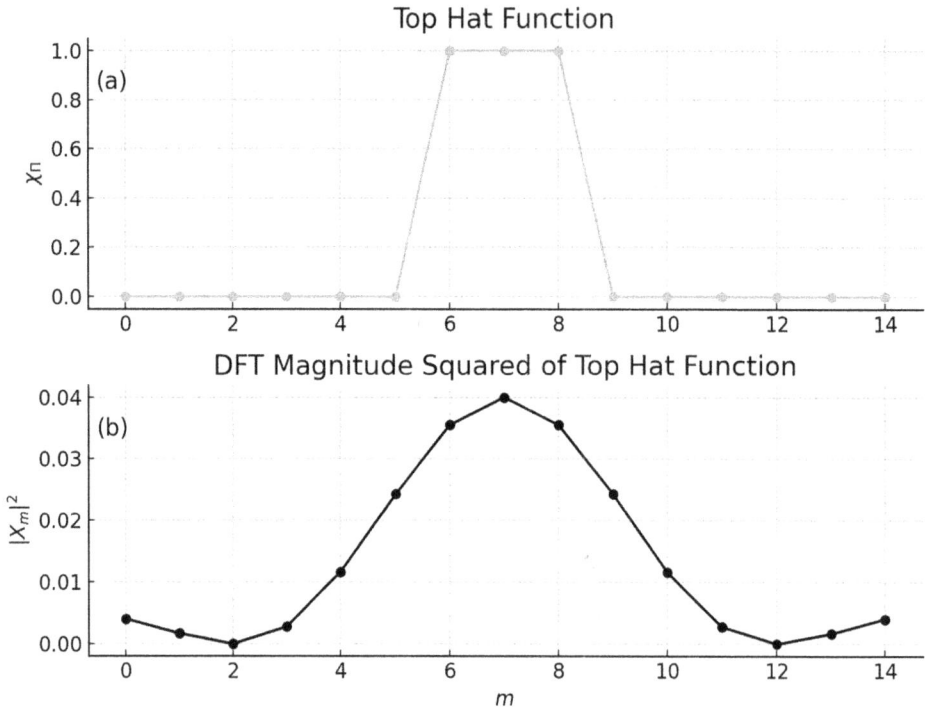

Figure 5.3. Digitized plots with connecting lines to make it easier to identify the shape. (a) The top hat function is oversampled at a ratio of 12/3. (b) The resolution of the DFT is significantly improved through oversampling, and the features of the sinc function are apparent.

sines and cosines. The larger the number of N terms, the more features can be sampled with the sinusoidal functions. We will revisit this phenomenon later using some visualizations.

5.4 Diffraction features

Now we know how to improve the resolution when calculating a diffraction; however, we can only see the central maximum and the first minimum in figure 5.3(b). How do we calculate more features? Let us add points to the top hat feature in our previous array (figure 5.4(a)). We also add some zero padding to ensure good resolution in the calculated diffraction pattern. With this change, the central maximum, the first maximum, and the first two minima are resolved.

How do the additional points in the diffraction feature (top hat) contribute to more features? The diffraction pattern relies on interference. Each additional point in the diffracting feature contributes an interference term $e^{2\pi in(m-(\frac{N-1}{2})/N)}$. The more interference terms we have, the more interference features can be resolved. We will visualize this phenomenon in the next section.

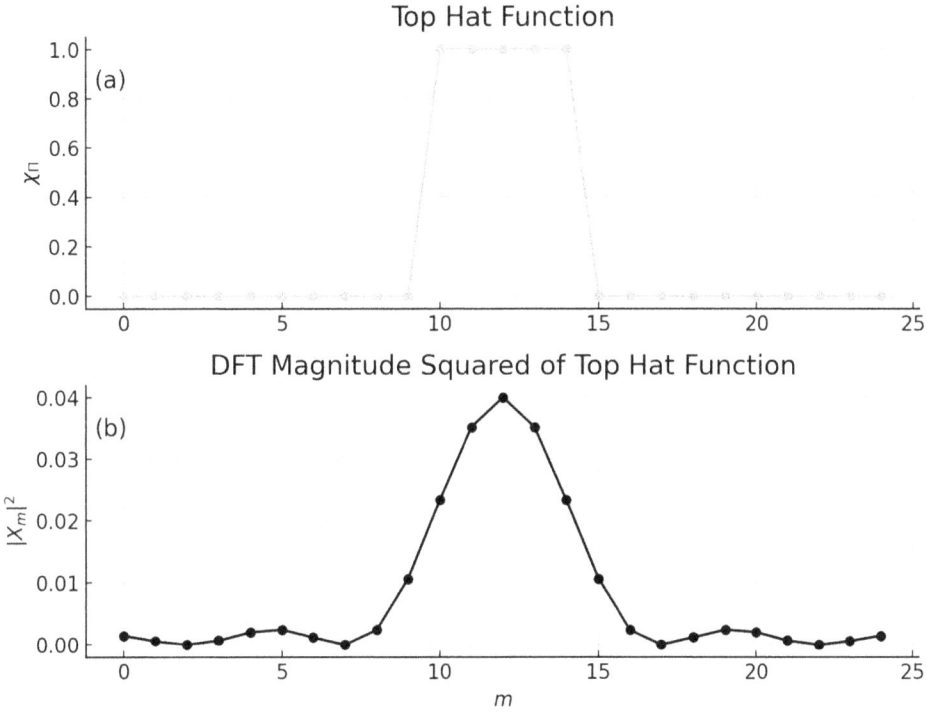

Figure 5.4. (a) The top hat function with five data points in the top hat feature. (b) DFT of the top hat function with the maximum and two minima resolved.

5.5 Matrix methods

In this section we will lay out DFTs using matrix methods. Matrix methods can provide a visual overview on how specific calculations function. Also, some computational software packages are matrix based. Understanding matrix methods will allow for smoother programming and avoid errors.

In matrix form, equation (5.3) can be written in index notation:

$$X_m = \frac{1}{N}\sum_{n=0}^{N-1} \chi_n \, e^{2\pi i n m/N} \quad \Leftrightarrow \quad \mathbf{X} = \frac{1}{N}[e^{2\pi i n m/N}]_{m,\,n=0}^{N-1}\chi, \tag{5.8}$$

where **X** is the DFT as a column vector and χ is the discretized aperture function in object space. Putting this expression into a matrix format:

$$\begin{bmatrix} X_0 \\ X_1 \\ \vdots \\ X_{N-1} \end{bmatrix} = \frac{1}{N} \begin{bmatrix} e^{2\pi i \cdot 0 \cdot 0/N} & e^{2\pi i \cdot 0 \cdot 1/N} & \cdots & e^{2\pi i \cdot 0 \cdot (N-1)/N} \\ e^{2\pi i \cdot 1 \cdot 0/N} & e^{2\pi i \cdot 1 \cdot 1/N} & \cdots & e^{2\pi i \cdot 1 \cdot (N-1)/N} \\ \vdots & \vdots & \ddots & \vdots \\ e^{2\pi i \cdot (N-1) \cdot 0/N} & e^{2\pi i \cdot (N-1) \cdot 1/N} & \cdots & e^{2\pi i \cdot (N-1) \cdot (N-1)/N} \end{bmatrix} \begin{bmatrix} \chi_0 \\ \chi_1 \\ \vdots \\ \chi_{N-1} \end{bmatrix} \tag{5.9}$$

We can rewrite equation (5.9) into a more compact and general form:

$$\mathbf{X} = \frac{1}{N} F \, \chi, \tag{5.10}$$

where F is the phase matrix. The phase matrix only depends on the size of the aperture function, not its content. Substituting $e^{2\pi i/N}$ with ω highlights symmetries in the phase matrix:

$$F = \begin{bmatrix} \omega^{0 \cdot 0} & \omega^{0 \cdot 1} & \cdots & \omega^{0 \cdot m} & \omega^{0 \cdot (N-1)} \\ \omega^{1 \cdot 0} & \omega^{1 \cdot 1} & \cdots & \omega^{1 \cdot m} & \omega^{1 \cdot (N-1)} \\ \vdots & \vdots & \ddots & \vdots & \\ \omega^{(N-1) \cdot 0} & \omega^{(N-1) \cdot 1} & \cdots & \omega^{(N-1) \cdot m} & \omega^{(N-1)(N-1)} \end{bmatrix} \tag{5.11}$$

Cleaning up the phase matrix a bit more and writing out a few more elements for clarity:

$$F = \begin{bmatrix} 1 & 1 & 1 & \cdots & 1 & 1 \\ 1 & \omega^{1} & \omega^{2} & \cdots & \omega^{1 \cdot m} & \omega^{1 \cdot (N-1)} \\ \vdots & \vdots & \ddots & \vdots & & \\ 1 & \omega^{(N-2) \cdot 1} & \omega^{(N-2) \cdot 2} & \cdots & \omega^{(N-2) \cdot m} & \omega^{(N-2)(N-1)} \\ 1 & \omega^{(N-1) \cdot 1} & \omega^{(N-1) \cdot 2} & \cdots & \omega^{(N-1) \cdot m} & \omega^{(N-1)(N-1)} \end{bmatrix} \tag{5.12}$$

The entire first row and first column are equal to one, since the exponents are always zeros because either n or m is equal to zero. This type of phase matrix is useful for frequency analysis of time-dependent phenomena.

5.5.1 Example: DFT $N = 4$

Let us calculate the DFT for $N = 4$ where $\chi = (0\ 0\ 1\ 1)$ using the matrix method, keeping in mind that the aperture function χ could be an edge.

Applying equation (5.10) we get:

$$\begin{bmatrix} X_0 \\ X_1 \\ X_2 \\ X_3 \end{bmatrix} = \frac{1}{4} \begin{bmatrix} 1 & 1 & 1 & 1 \\ 1 & i & -1 & -i \\ 1 & -1 & 1 & -1 \\ 1 & -i & -1 & i \end{bmatrix} \begin{bmatrix} 0 \\ 0 \\ 1 \\ 1 \end{bmatrix} \tag{5.13}$$

We notice that there are some symmetries in the phase matrix. We can see how symmetries in the phase matrix can be exploited to shorten computing time. Next matrix multiplication reveals:

$$\mathbf{X} = \frac{1}{4} \begin{bmatrix} 1 + 1 \\ -1 + (-i) \\ 1 + (-1) \\ -1 + i \end{bmatrix} = \frac{1}{4} \begin{bmatrix} 2 \\ -1 - i \\ 0 \\ -1 + i \end{bmatrix} \tag{5.14}$$

For diffraction it makes sense to center the diffraction pattern as we did in figure 5.2(b) by centering the phase matrix. This will center the diffraction pattern.

$$
F = \begin{bmatrix}
\omega^{\left(-\frac{N-1}{2}\right)\left(-\frac{N-1}{2}\right)} & \cdots & \omega^{\left(-\frac{N-1}{2}\right)\cdot 0} & \cdots & \omega^{\left(-\frac{N-1}{2}\right)\left(\frac{N-1}{2}\right)} \\
\vdots & \ddots & \vdots & \ddots & \vdots \\
\omega^{0 \cdot \left(-\frac{N-1}{2}\right)} & \cdots & \omega^{0 \cdot 0} & \cdots & \omega^{0 \cdot \left(\frac{N-1}{2}\right)} \\
\vdots & \ddots & \vdots & \ddots & \vdots \\
\omega^{\left(\frac{N-1}{2}\right)\left(-\frac{N-1}{2}\right)} & \cdots & \omega^{\left(\frac{N-1}{2}\right)\cdot 0} & \cdots & \omega^{\left(\frac{N-1}{2}\right)\left(\frac{N-1}{2}\right)}
\end{bmatrix}
\tag{5.15}
$$

Cleaning up the matrix to emphasize the centering of the ones:

$$
F = \begin{bmatrix}
\omega^{\left(-\frac{N-1}{2}\right)\left(-\frac{N-1}{2}\right)} & \cdots & 1 & \cdots & \omega^{\left(-\frac{N-1}{2}\right)\left(\frac{N-1}{2}\right)} \\
\vdots & \ddots & \vdots & \ddots & \vdots \\
1 & \cdots & 1 & \cdots & 1 \\
\vdots & \ddots & \vdots & \ddots & \vdots \\
\omega^{\left(\frac{N-1}{2}\right)\left(-\frac{N-1}{2}\right)} & \cdots & 1 & \cdots & \omega^{\left(\frac{N-1}{2}\right)\left(\frac{N-1}{2}\right)}
\end{bmatrix}.
\tag{5.16}
$$

In summary, we did not rename the phase matrix F when we centered the column and row ones simply because they are both valid phase matrices. Going forward we will use the centered phase matrix and call it F. For even N the phase matrix can be written.

$$
F = \begin{bmatrix}
\omega^{\left(-\frac{N}{2}\right)\left(-\frac{N}{2}\right)} & \cdots & 1 & \cdots & \omega^{\left(-\frac{N}{2}\right)\left(\frac{N}{2}\right)} \\
\vdots & \ddots & \vdots & \ddots & \vdots \\
1 & \cdots & 1 & \cdots & 1 \\
\vdots & \ddots & \vdots & \ddots & \vdots \\
\omega^{\left(\frac{N}{2}\right)\left(-\frac{N}{2}\right)} & \cdots & 1 & \cdots & \omega^{\left(\frac{N}{2}\right)\left(\frac{N}{2}\right)}.
\end{bmatrix}.
\tag{5.17}
$$

One may argue that it is impossible to center a phase matrix for even N. In the typical computation $N > 100$ or so. In that case, the off center is negligible and not noticeable.

5.6 Visual interpretation

The phase matrix does not depend on the aperture function; it provides a grid that is related to the aperture function via matrix multiplication. Figure 5.5 shows the real and imaginary parts of the phase matrix. The points indicate the values of the phase matrix. The yellow lines are merely estimated fits, so they may be imperfect, as this is inherent in the nature of discretized computations. The points in between the matrix elements are often missing or interpolated.

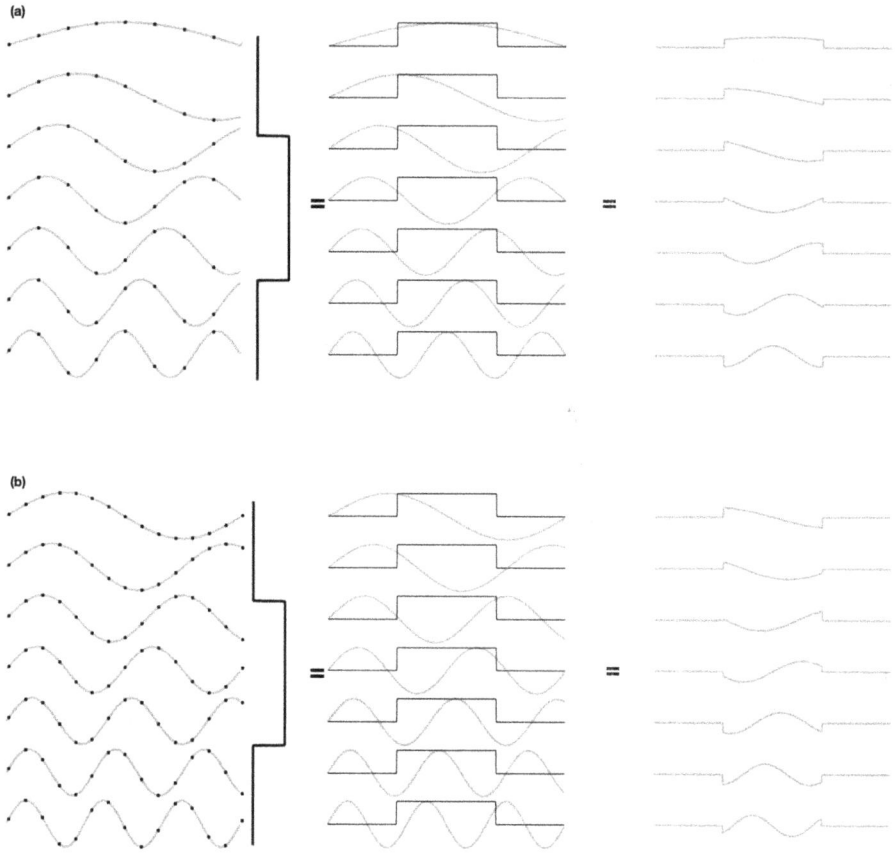

Figure 5.5. A visual representation of the phase matrix multiplying the top hat function. The graphs show the real (a) and imaginary (b) parts of the 9×9 phase matrix with the top hat function to the right representing a column vector. The overlay on the right represents the multiplication of the top hat function with each row of the phase matrix.

Figure 5.5(a) illustrates the real part of matrix multiplication using cosines. Notice that the symmetries of cosines match the symmetry of the top hat function. They are both even functions. When an even function is multiplied by an even function, the result is likely to be positive values. Adding these values gives a net positive result.

In contrast, figure 5.5(b) illustrates the imaginary part of the matrix multiplication using sines. Notice that the symmetries of sines do not match the symmetry of the top-hat function. The sine is an odd function, while the top-hat function is positioned as an even function. When an odd and an even function are multiplied, they will result in the same number of even and odd numbers. Adding these values gives a net result. In the same way, integrating over one period of a sinusoid yields a zero value.

Taking into account the symmetry considerations in figure 5.5, we can see that the phase matrix not only probes the frequencies present in the diffracting object (aperture function) but also takes symmetries into account. The DFT of a symmetric aperture function is represented by real numbers only, while imaginary numbers represent an asymmetric aperture function. The real part in a DFT consists of a series of cosine functions, while the imaginary part consists of sine functions only. Of course, a mix of both symmetric and asymmetric properties is also possible.

Figure 5.5 also helps us understand the resolution, which was introduced in section 5.3. The more points in the phase matrix, the more frequencies will be probed. We can see that by considering the phase matrix exponent, which is the phase. For an even number of elements N, the phase magnitude ranges from 0 to $2\pi i(N/2)^2/N = iN\pi/2$. In other words, the more matrix elements, the higher the frequencies that are included. This allows for better resolution, especially for a structure like the top-hat function, which has sharp edges that can be defined more accurately by including higher frequencies.

Similarly, diffraction features, which were introduced in section 5.4, can be probed in more detail with more points. Probing only three points (Figure 5.3) allows for a reference point that can show a destructive and a constructive interference point; however, using five points to probe the structure allows for an additional two interference points, a maximum and a minimum (figure 5.4).

5.6.1 Practice problem

Use equation (5.10) to calculate the DFT of $\chi_\Pi = (1\ 1\ 0\ 0)$. Be sure to write out the phase matrix. Note the real and imaginary values. What do you notice in the phase matrix and the results? How does this problem differ from the example in section 5.5.1?

5.7 Fast Fourier transform (FFT)

In this section, we outline the foundation and purpose of a fast Fourier transform (FFT). We will explain only what is necessary to understand the benefits and uses of an FFT and why it is used in computing.

The example in section 5.5.1 and the practice problem in section 5.6.1 illustrate numerous symmetries in the phase matrix. These symmetries help reflect symmetries in the DFT, and, based on the symmetries in the phase matrix, we could save ourselves calculation time and, if automated, computing time and resources.

Figure 5.6 illustrates the symmetries in various ways. The diagonal symmetries in the phases are evident in figure 5.6(a). Diagonal symmetries alone can reduce calculations by 1/2. Additionally, each quarter exhibits diagonal symmetries, again reducing the number of calculations. The real part (Figure 5.6(b)) shows that each quarter contains identical values, while the imaginary elements (figure 5.6(c)) exhibit an odd symmetry. We might also consider that the square of any real matrix element added to the square of any imaginary element must be equal to 1 since $\sin^2 \phi_{mn} + \cos^2 \phi_{mn} = 1$.

In summary, it takes $O(N^2)$ calculations to compute a DFT. Using an FFT reduces the requirement to $O(N \log_2 N)$ calculations [4]. To illustrate, an array of

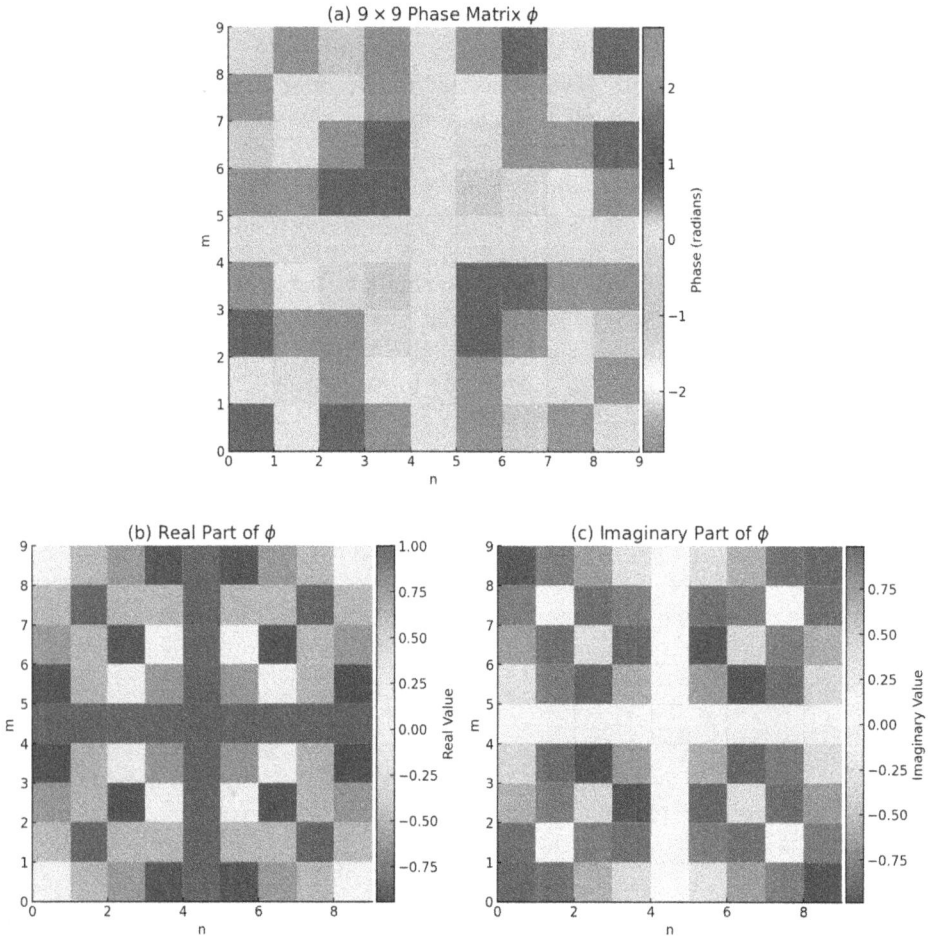

Figure 5.6. Color-coded 9×9 phase matrix representations as they appear in equation (5.23). (a) Phases (exponents), (b) real parts, and (c) imaginary parts of the matrix elements.

512 elements would require $512 \times 512 = 262\ 144$ calculations using a DFT, but only $512 \log_2 512 = 512 \times 9 = 4608$ calculations using an FFT.

5.8 Two-dimensional Fourier transform

Now that we have reviewed the principle of DFTs, we are moving towards more practical applications. Most physical phenomena take place in three dimensions; however, it turns out that for the type of diffraction we are dealing with, two dimensions suffice. We derived the analytical Fourier transform equation (4.12) for a one-dimensional aperture function $g(x)$. We can expand that expression to the two-dimensional case if we imagine a two-dimensional aperture function $g(x, y)$ such as a circle or rectangle at the same distance from the screen as the one-dimensional aperture function and parallel to the screen. We can write the two-dimensional Fourier integral:

$$E(x, y) = E_0 \int_{x=-\infty}^{x=+\infty} \int_{y=-\infty}^{y=+\infty} g(x, y)e^{i(k_x x + k_y y)} dx dy. \tag{5.18}$$

Recognizing that $(k_x x + k_y y) = \boldsymbol{k} \cdot \boldsymbol{r}$, we can even write a general expression of the analytic FT in three dimensions:

$$E(x, y, z) = E_0 \iiint_{\mathbb{R}^3} g(x, y)e^{i\boldsymbol{k} \cdot \boldsymbol{r}} d^3\boldsymbol{r}, \tag{5.19}$$

where $d^3\boldsymbol{r} = dx dy dz$ and the integral is over all space. However, for our purposes, $z \gg x, z \gg y$, and $z \approx r$. The effects of phase shifts in the z-direction are negligible. As a result, we can just focus on the two-dimensional case, described by equation (5.18).

5.9 Two-dimensional discrete Fourier transform

Of course, now that there is an analytical two-dimensional FT, we can adapt equation (5.7) to two dimensions:

$$X_{m_x, m_y} = \frac{1}{N_x N_y} \sum_{n_x = -\frac{N_x-1}{2}}^{\frac{N_x-1}{2}} \sum_{n_y = -\frac{N_y-1}{2}}^{\frac{N_y-1}{2}} X_{n_x, n_y} e^{2\pi i \left(\frac{n_x m_x}{N_x} + \frac{n_y m_y}{N_y} \right)} \quad \text{for odd } N_x, N_y$$

$$\tag{5.20}$$

$$X_{m_x, m_y} = \frac{1}{N_x N_y} \sum_{n_x = -\frac{N_x}{2}}^{\frac{N_x}{2}} \sum_{n_y = -\frac{N_y}{2}}^{\frac{N_y}{2}} X_{n_x, n_y} e^{2\pi i \left(\frac{n_x m_x}{N_x} + \frac{n_y m_y}{N_y} \right)} \quad \text{for even } N_x, N_y$$

We can see that even and odd cases for N_x and N_y can also be mixed; however, in many cases we choose $N_x = N_y$ so that equation (5.20) becomes:

$$X_{m_x, m_y} = \frac{1}{N^2} \sum_{n_x = -\frac{N-1}{2}}^{\frac{N-1}{2}} \sum_{n_y = -\frac{N-1}{2}}^{\frac{N-1}{2}} X_{n_x, n_y} e^{2\pi i \left(\frac{n_x m_x}{N} + \frac{n_y m_y}{N} \right)} \quad \text{for odd } N$$

$$\tag{5.21}$$

$$X_{m_x, m_y} = \frac{1}{N^2} \sum_{n_x = -\frac{N}{2}}^{\frac{N}{2}} \sum_{n_y = -\frac{N}{2}}^{\frac{N}{2}} X_{n_x, n_y} e^{2\pi i \left(\frac{n_x m_x}{N} + \frac{n_y m_y}{N} \right)} \quad \text{for even } N$$

5.10 Matrix format in two dimensions

We saw in section 5.5 that matrices can lead to some compelling visualizations, as we also observed in section 5.6. In a two-dimensional matrix format, equation (5.21) can be written as follows:

$$\mathbf{X} = \frac{1}{N^2} F \chi F^\dagger, \tag{5.22}$$

where $F^\dagger$ is the hermitian conjugate of the phase matrix F; that is the transpose and complex conjugate of F. Writing out some of the matrix elements $\omega^{n_x m_x}$ for odd N, the phase matrix F looks like:

$$
F = \begin{bmatrix}
\omega^{(-\frac{N-1}{2})(-\frac{N-1}{2})} & \cdots & \omega^{(-\frac{N-1}{2})(-1)} & \omega^{(-\frac{N-1}{2})(0)} & \cdots & \omega^{(-\frac{N-1}{2})(\frac{N-1}{2})} \\
\vdots & \ddots & \vdots & \vdots & \ddots & \vdots \\
\omega^{0\cdot(-\frac{N-1}{2})} & \cdots & \omega^{0\cdot(-1)} & \omega^{0\cdot0} & \cdots & \omega^{0\cdot(\frac{N-1}{2})} \\
\vdots & \ddots & \vdots & \vdots & \ddots & \vdots \\
\omega^{(\frac{N-1}{2})(-\frac{N-1}{2})} & \cdots & \omega^{(\frac{N-1}{2})(-1)} & \omega^{(\frac{N-1}{2})(0)} & \cdots & \omega^{(\frac{N-1}{2})(\frac{N-1}{2})}
\end{bmatrix}, \tag{5.23}
$$

which can be simplified:

$$
F = \begin{bmatrix}
\omega^{(-\frac{N-1}{2})^2} & \cdots & \omega^{\frac{N-1}{2}} & 1 & \cdots & \omega^{-(\frac{N-1}{2})^2} \\
\vdots & \ddots & \vdots & \vdots & \ddots & \vdots \\
1 & \cdots & 1 & 1 & \cdots & 1 \\
\vdots & \ddots & \vdots & \vdots & \ddots & \vdots \\
\omega^{-(\frac{N-1}{2})^2} & \cdots & \omega^{-\frac{N-1}{2}} & 1 & \cdots & \omega^{(\frac{N-1}{2})^2}
\end{bmatrix}. \tag{5.24}
$$

We will leave it to the reader to generate the phase matrix for even N. In general, the matrix multiplication using equation (5.22) becomes:

$$
\mathbf{X} = \frac{1}{N^2} \begin{bmatrix}
\omega^{n_x m_x} & \cdots & \omega^{n_x m_x} \\
\vdots & \ddots & \vdots \\
\omega^{n_x m_x} & \cdots & \omega^{n_x m_x}
\end{bmatrix}
\begin{bmatrix}
\chi_{m_x, m_y} & \cdots & \chi_{m_x, m_y} \\
\vdots & \ddots & \vdots \\
\chi_{m_x, m_y} & \cdots & \chi_{m_x, m_y}
\end{bmatrix}
\begin{bmatrix}
\omega^{n_y m_y} & \cdots & \omega^{n_y m_y} \\
\vdots & \ddots & \vdots \\
\omega^{n_y m_y} & \cdots & \omega^{n_y m_y}
\end{bmatrix}. \tag{5.25}
$$

The phase matrix F on the left (equation (5.25)) probes the columns in the aperture matrix in the middle, while the hermitian conjugate $F^\dagger$ on the right probes the rows of the aperture matrix, as we saw in the one-dimensional case using figure 5.5. In this case, $F = F^\dagger$, since we limit ourselves to a square $N \times N$ aperture matrix.

5.10.1 Example: DFT $N = 9$

Let us use a two-dimensional version of the top-hat function as an aperture function and calculate the DFT, followed by the diffraction pattern (power spectrum). We will use a square:

$$
\chi = \begin{bmatrix}
0 & 0 & 0 & 0 & 0 & 0 & 0 & 0 & 0 \\
0 & 0 & 0 & 0 & 0 & 0 & 0 & 0 & 0 \\
0 & 0 & 0 & 0 & 0 & 0 & 0 & 0 & 0 \\
0 & 0 & 0 & 1 & 1 & 1 & 0 & 0 & 0 \\
0 & 0 & 0 & 1 & 1 & 1 & 0 & 0 & 0 \\
0 & 0 & 0 & 1 & 1 & 1 & 0 & 0 & 0 \\
0 & 0 & 0 & 0 & 0 & 0 & 0 & 0 & 0 \\
0 & 0 & 0 & 0 & 0 & 0 & 0 & 0 & 0 \\
0 & 0 & 0 & 0 & 0 & 0 & 0 & 0 & 0
\end{bmatrix}
$$

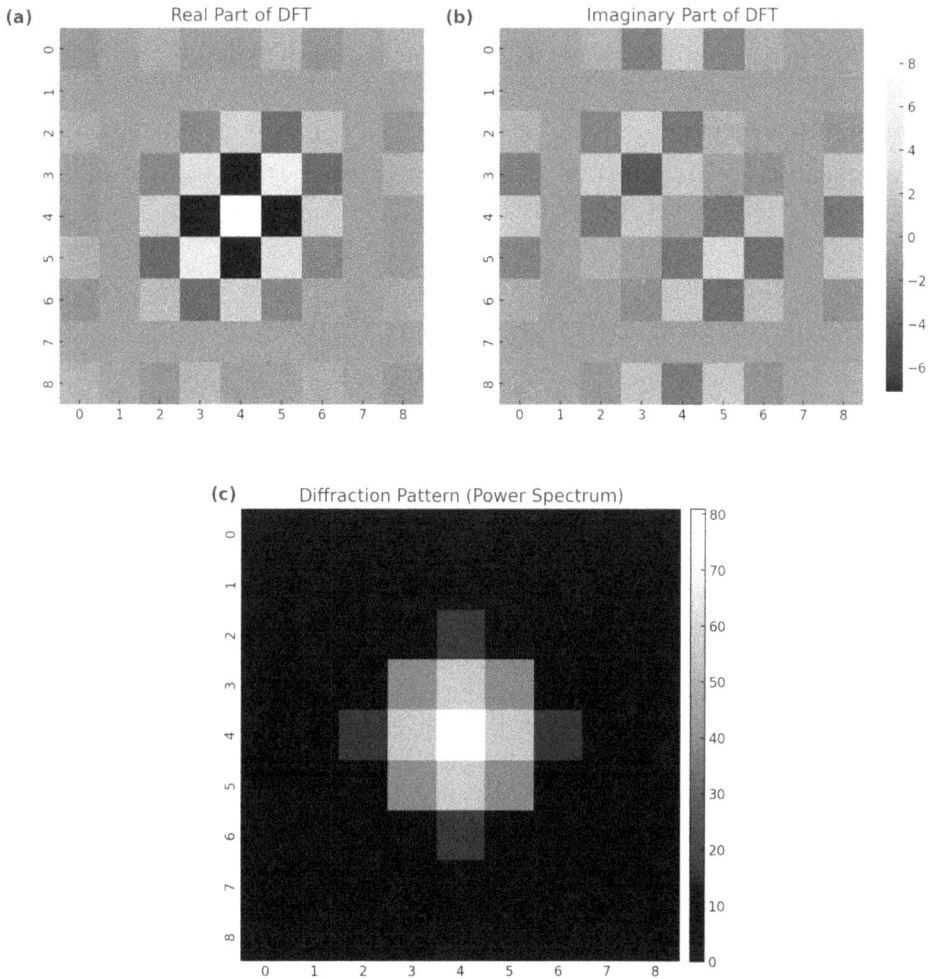

Figure 5.7. (a) Real and (b) imaginary part of the DFT of a 3×3 square in a 9×9 matrix and the corresponding (c) calculated diffraction pattern.

We will leave detailed calculations for an exercise; however, plotting the real and imaginary parts of the DFT as weel as the power spectrum $|\chi|^2$ as a heat map shows what can be expected (figure 5.7): grainy outlines. We expected only real parts due to the symmetry of the two-dimensional structure; however, figure 5.7(b) shows some imaginary values closer to zero because of the discrete nature of the DFT. As we will see more of later, we often have to take into account that computations are discrete in nature and the numerical values are finite leading to rounding errors.

The diffraction pattern itself matches the symmetry of a square figure 5.7(c). The cross-section in the horizontal and vertical directions matches a sinc function as the aperture in each direction is that of a one-dimensional single slit, which we discussed in sections 4.3.1 and 5.2. As discussed earlier, the graininess is due to a lack of oversampling; i.e., a lack of zeros around the square, while the diffraction features are lacking due to a grainy sampling of the square, which is the diffracting feature.

5.10.2 Practice problem

Using equation (5.22), calculate the two-dimensional DFT and the diffraction pattern of the following 4×4 matrix:

$$\chi = \begin{bmatrix} 1 1 1 1 \\ 1 0 0 1 \\ 1 0 0 1 \\ 1 1 1 1 \end{bmatrix}$$

Create three heat maps: one for real and imaginary values, and another for the diffraction pattern. Does your result make sense? Why? What do you notice?

5.11 Computational practicalities

We established in section 5.3 that the resolution of a calculated diffraction pattern can be controlled by oversampling. In addition, section 5.4 established that the extent of the interference features is governed by how many frequencies sample the diffracting object. So far, we have limited the number of matrix elements to allow for manual computation of diffraction patterns. Of course, the primary advantage of using computational methods lies in the convenience of large-scale computations and those that are not analytically accessible. Here, we will explore two-dimensional examples as they are the most applicable.

We encourage the reader to reproduce the computations using their own favorite software, such as Mathematica, MatLab, or Python, while adjusting various parameters, such as the aperture function or the size of the matrix.

5.11.1 Power spectrum considerations

The power spectrum is directly proportional to the diffraction pattern as indicated by equation (4.13). More generally, the power spectrum is given by the following:

$$P = |\mathbf{X}|^2 = \mathbf{X}^*\mathbf{X}. \tag{5.26}$$

Of course, we already learned that the light intensity I, which is what we measure in a diffraction pattern, is proportional to the electric field amplitude squared. Let us compute the power spectrum of a 32×32 matrix with an embedded square (figure 5.8 using an FFT to save time. Choosing 2^n dimensions for our matrix and features when possible takes advantage of the binary nature of a computer and reduces computing time.

Figure 5.9 shows several versions of an appropriately grainy power spectrum represented as heat maps. The interference effect of the diffraction is not visible in figure 5.9 (a). The intensities are plotted on a linear scale; however, we can see interference patterns with the naked eye since eyes are nonlinear sensors. Even common cameras have a nonlinear dynamic range to adapt to light levels and avoid

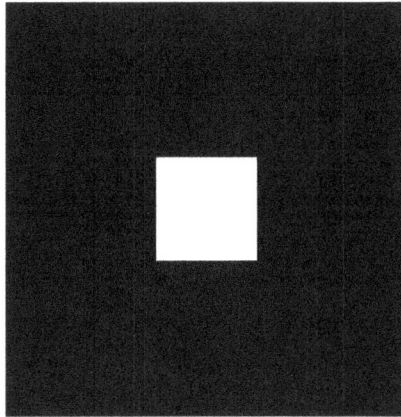

Figure 5.8. A representative image of a 32×32 matrix with an 8×8 square of ones in white in the center framed by zeros in black.

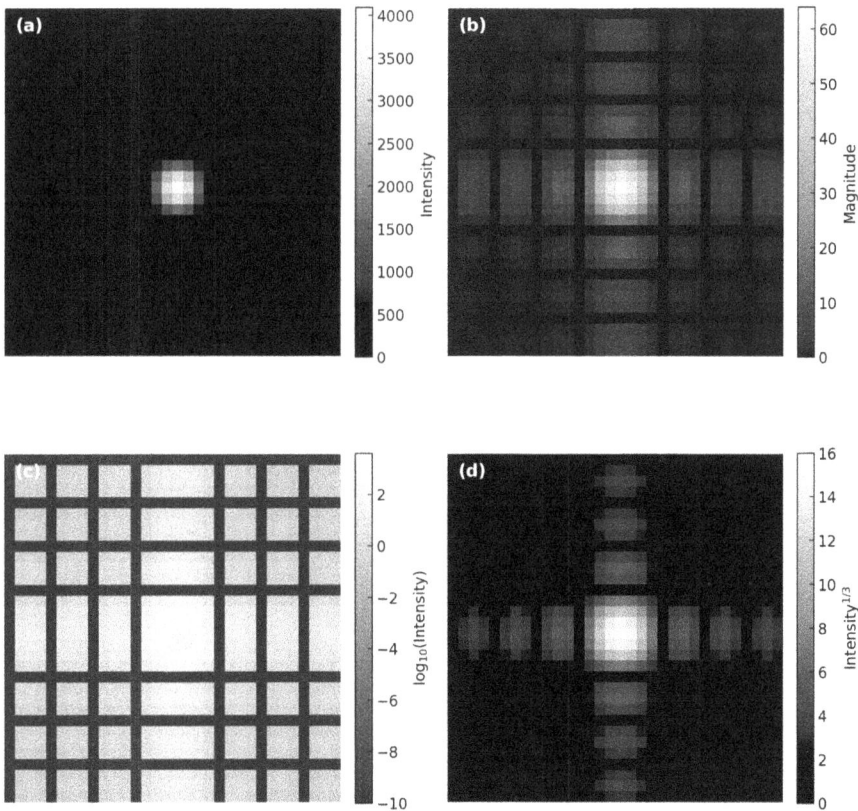

Figure 5.9. Diffraction pattern of a 32×32 matrix with an 8×8 aperture. Heat map of the (a) intensity (power spectrum) $P = |\mathbf{X}|^2$. The interference effects are not visible on this linear scale. Heat maps of the (b) magnitude $\sqrt{P} = |\mathbf{X}|$, (c) log base 10 of the intensity, and the (d) cube root of the intensity.

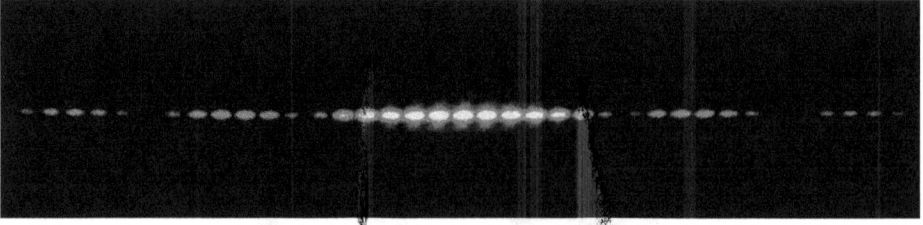

Figure 5.10. Photograph of an experimental diffraction pattern due to a double slit using 650 nm (red) laser light from laser diode. Credit: Merlo Lab 2025.

saturation. Figures 5.9(b)–(d) show various nonlinear scales as plotting tools to examine power spectra across scales.

Figure 5.10 shows the double-slit diffraction pattern due to red light. The white regions demonstrate camera saturation; however, the interference features are clearly distinguishable.

In summary, computed diffraction patterns often have to be viewed on a nonlinear scale to examine diffraction features. This can be accomplished by scaling the power spectrum using a logarithm or fractional exponent, such as a square root.

5.11.2 Resolution revisited

Here we explore the computational limits of resolution. In section 5.3, we discussed that oversampling is critical to resolving a diffraction pattern. Our example in the previous section involves a 32×32 matrix with an 8×8 aperture. We can determine the oversampling ratio:

$$M = \frac{f_s}{f_{\text{Nyquist}}}, \tag{5.27}$$

where f_s is the Fourier domain sampling frequency and f_{Nyquist} is the Nyquist frequency, which is determined by the size of the object. In the one-dimensional case, the oversampling ratio M must be at least 2 to obtain a resolution that resolves some diffraction characteristics. For the two-dimensional case $2^{1/2}$ is the minimum oversampling bound [5]. Notice that the oversampling ratio has many applications in the spatial and time domains. Here, we focus on the spatial domain since we are dealing with images.

Alternatively, the oversampling ratio can be defined by area (or in a higher dimension by volume) so that the oversampling ratio σ must be minimally 2:

$$\sigma = \frac{\text{total area}}{\text{aperture area}}. \tag{5.28}$$

Figure 5.11 shows how much the resolution improves by embedding the 8×8 square in a 64×64 matrix, increasing the oversampling ratio σ to 64. It is clear from some of the above examples, such as the 9×9 matrix in figure 5.7, that minimum

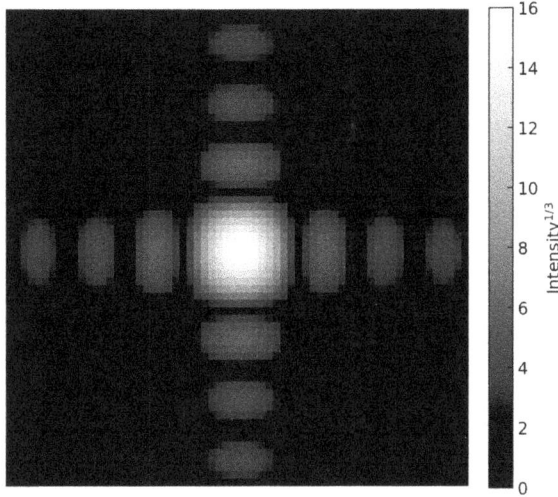

Figure 5.11. Recomputed diffraction pattern for the 8×8 aperture in figure 5.8 embedded in a 64×64 matrix.

oversampling is often not enough to distinguish features. The desired resolution frequently depends on the application.

5.11.3 Example: oversampling limit for $N = 9$

Let us calculate the oversampling ratio for the example in section 5.10.1 using equation (5.27). We need to calculate the sampling frequency f_s and the Nyquist frequency f_{Nyquist}. The Fourier domain frequency for a spatial case is given by:

$$f_s = 1/\Delta x, \tag{5.29}$$

where

$$f_s = FOV/N. \tag{5.30}$$

The FOV (field of view) is the extent of the image in one dimension, and N is the number of elements in that dimension. In this case, the FOV is expressed in matrix elements (9) and $N = 9$ so that $f_s = 1$. The Nyquist frequency is

$$f_{\text{Nyquist}} = 1/(2 \cdot \text{aperture size}). \tag{5.31}$$

In the example in section 5.10.1 the object is three matrix elements wide, so $f_{\text{Nyquist}} = 1/(2 \cdot 3)$.

Combining the above results in the oversampling ratio $M = \frac{1}{1/6} = 6$.

We conclude $M > \sqrt{2}$. Note that the Nyquist frequency guides the minimum oversampling ratio. It is always advisable to strive for a larger oversampling ratio. Of course, it must be balanced with a reasonable computing time. A huge oversampling ratio will result in unreasonably large computing times.

5.11.4 Practice problem

Determine the oversampling ratio for the practice problem in section 5.10.2. Do you need to adjust the oversampling ratio to meet the minimum criterion for resolving the diffraction pattern? If so, how would you change the matrix?

5.12 Aliasing

In the previous sections, we saw that a diffraction pattern may appear 'grainy' when there are not enough matrix elements. This is a form of aliasing. The image will appear smoother when we increase the oversampling ratio. In figure 5.11, the larger oversampling ratio of 64 smooths the diffraction image considerably. In figure 5.12, we increased the square size to 16×16 matrix elements to capture more diffraction features. We also increased the total matrix size to 2048×2048 to achieve an oversampling ratio σ of 16 384. The diffraction pattern appears continuous now.

Let us now increase the aperture size to 64×64 to capture more diffraction features and use a 1024×1024 matrix to save computing time. The oversampling ratio in figure 5.13, σ is now reduced to 256. The diffraction features are exaggerated but plausible.

Using the same 1024×1024 matrix but replacing the square with a circular structure, we expect concentric diffraction rings with a sinc-function as a cross-section. The oversampling ratio is now larger than that for the square; however, we see some unexpected structures towards the edges (figure 5.14).

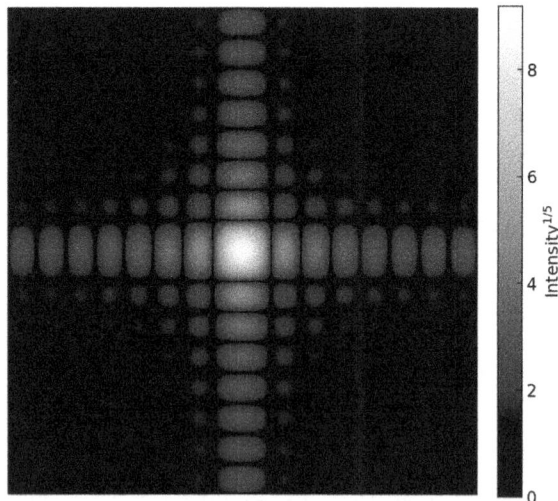

Figure 5.12. Recomputed diffraction pattern for the 16×16 square aperture embedded in a 2048×2048 matrix.

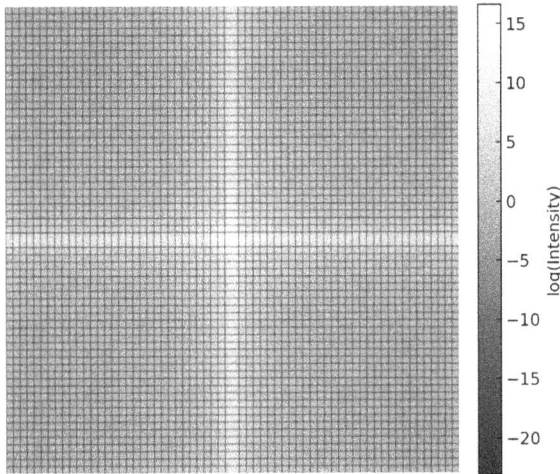

Figure 5.13. Recomputed diffraction pattern for the 64 × 64 square aperture embedded in a 1024 × 1024 matrix.

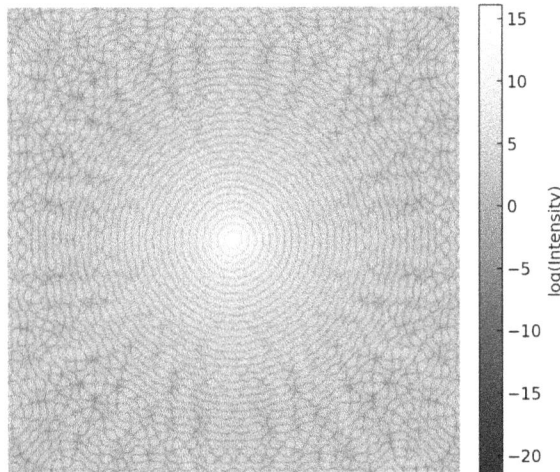

Figure 5.14. Recomputed diffraction pattern for the circular aperture with a diameter of 64 matrix elements embedded in a 1024 × 1024 matrix.

5.13 Conclusion

In this chapter, we learned how to calculate Fourier transforms using digital means. DFTs have many advantages. In this text, we focus on diffraction patterns; however, many fields such as image processing, signal processing, quantum mechanics, and many other fields take advantage of DFTs because they can easily be automated. We saw in the previous chapter that analytical Fourier transforms can help with conceptual understandings and certain types of problems; nevertheless, they can be tedious or solutions might not be accessible via analytical calculations.

References

[1] Gauß C F 1877 *Theoria Attractionis Corporum Sphaeroidicorum Ellipticorum Homogeneorum: Methodo Nova Tractata* (Berlin: Springer)

[2] Fourier J -B J 1807 *I. Théorie de la propagation de la chaleur dans les solides* (Paris: Institute of France)

[3] Miao J, Ishikawa T, Anderson E H and Hodgson K O 2003 Phase retrieval of diffraction patterns from noncrystalline samples using the oversampling method *Phys. Rev.* B **67** 174104

[4] Brigham E O 1988 *The Fast Fourier Transform and Its Applications* (Hoboken, NJ: Prentice-Hall)

[5] Miao J, Sayre D and Chapman H N 1998 Phase retrieval from the magnitude of the Fourier transforms of nonperiodic objects *J. Opt. Soc. Am.* A **15** 1662–9

IOP Publishing

Optical Interference and Dynamic Diffraction
Research methods for undergraduates
Jenny Magnes and Juan M Merlo-Ramírez

Chapter 6

Dynamic diffraction

6.1 Introduction

This chapter brings together previous chapters and explores the diffraction patterns of moving objects. The extraction of useful information and modeling is explained. How does dynamic optical diffraction (DOD) contribute to understanding dynamic systems?

6.2 Dynamic single slit

Let us model how the diffraction pattern of a slit as shown in figure 4.4 is affected by a change in width $a(t)$ so that equation (4.20) becomes:

$$I(k_x, t) = E_0^2 a(t)^2 \operatorname{sinc}^2\left(\frac{k_x a(t)}{2}\right). \tag{6.1}$$

Now, we can apply various time-dependent versions of $a(t)$ to explore the effects in the diffraction pattern.

6.2.1 Linear changes

Let us assume that $a(t)$ changes linearly with time so that $a(t) = C \cdot t$, where C is a constant. Equation (6.7) becomes:

$$I(k_x, t) = E_0^2 C^2 \cdot t^2 \operatorname{sinc}^2\left(\frac{k_x C \cdot t}{2}\right). \tag{6.2}$$

We can pick convenient values for the constants E_0 and C, to study the time evolution of the diffraction pattern (figure 6.1). Letting E_0 and C equal 1, the following expression will suffice in studying the diffraction pattern taking snap shots at different points in time:

doi:10.1088/978-0-7503-4836-2ch6

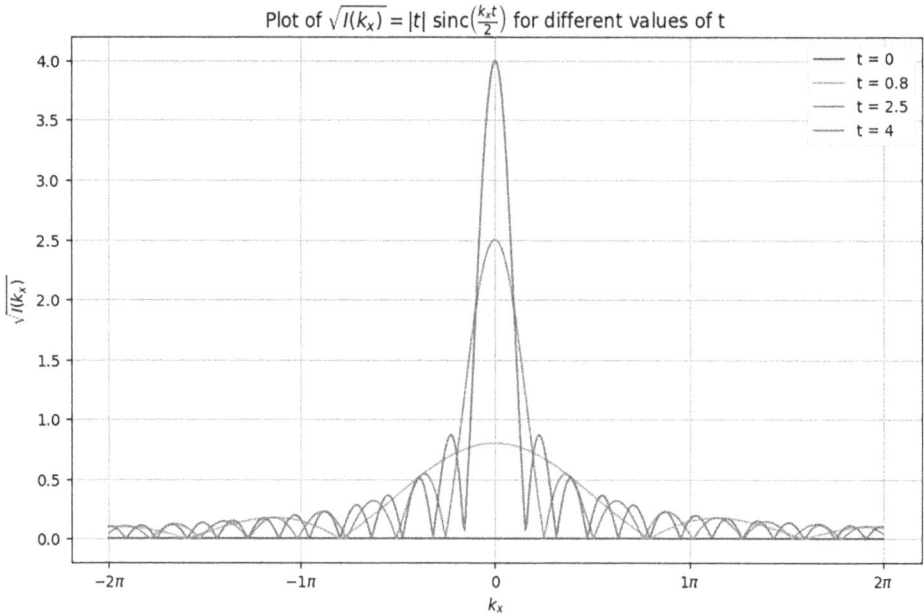

Figure 6.1. The diffraction pattern due to a single slit with an oscillating slit width $a(t)$. Setting the constants E_0 and C equal to 1, the diffraction pattern shows an increasing interference structure as the slit widens at a linear rate C. The intensity is scaled to the square root (the magnitude of the electric field) to make small fluctuations visible, as discussed in section 5.11.1.

$$I(k_x) = t^2 \operatorname{sinc}^2\left(\frac{k_x t}{2}\right). \tag{6.3}$$

We see the narrowing of the diffraction pattern as the slit widens and time evolves. This is the familiar behavior of reciprocal space in Fourier pairs. Note that at $t = 0$, the slit width is equal to zero, so there is no diffraction pattern. We can also say that the slit is infinitely narrow, so that the diffraction pattern is infinitely wide, as can be seen in figure 6.1.

Now, we place a detector at a stationary point, a particular k_x, in the diffraction pattern, such as a photodiode (PD). Keeping k_x constant, the PD detects interference fluctuations as the slit widens. Holding k_x constant, equation (6.3) becomes:

$$I(t) = \frac{4}{k_x^2} \sin^2\left(\frac{k_x t}{2}\right). \tag{6.4}$$

We can verify the validity of this graph by examining the periods of the fluctuations. The sine function should be equal to zero whenever the phase is equal to an integer multiple of π. The phase $\frac{k_x t}{2} = 2\pi f$, where f is the frequency of a sine function; however, in this case f is only half of the frequency F of the curve in figure 6.2 since the sine function is squared, doubling the frequency. $\frac{k_x t}{2} = \pi F$. As a result, the frequency at which the intensity $I(t)$ oscillates is dependent on the

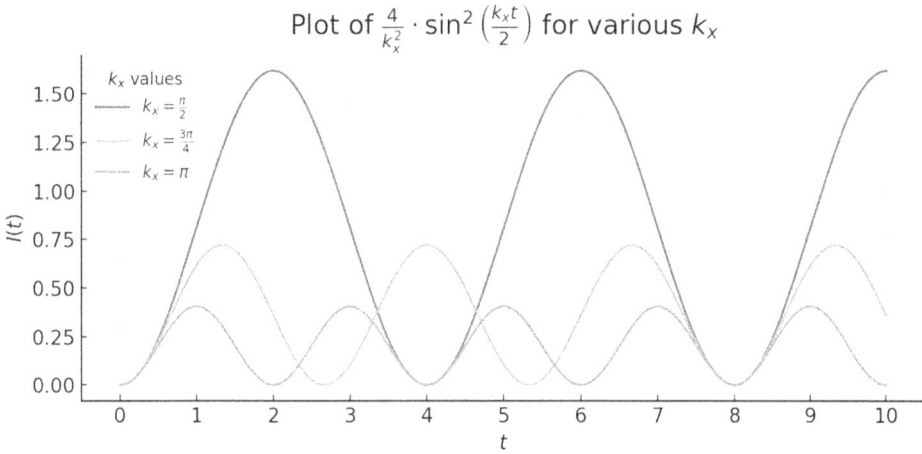

Figure 6.2. The intensity $I(t)$ at various points in the diffraction patter. The frequency of the power fluctuations increases with k_x as the point sensor moves away from the central maximum.

Table 6.1. Period T for different values of k_x. The period decreases as the distance from the central maximum increases.

k_x	$T = \frac{2\pi}{k_x}$
$\frac{\pi}{2}$	4.0
$\frac{3\pi}{4}$	2.7
π	2.0

location in the diffraction pattern. The location is tied to $k_x = k\frac{x_s}{d}$ as examined in equation (4.9), where x_s is the distance from the central maximum on the screen and d is the distance between the slit and the screen. $F = \frac{k_x}{2\pi}$. Table 6.1 shows the different periods $T = 1/F$ as they appear in figure 6.2.

The calculated periods in the table match the periods in the graph. The period T decreases, with increasing distance related to k_x, from the central maximum. Eventually, the period would be too small to measure. Most importantly, these calculations are all derived from approximations; i.e., the distance from the slit to the screen is much smaller than the width of the slit, and the angle from the central maximum is small, meaning that k_x is limited to relatively small values. This makes physical sense as the light would fade out for very large angles, and a very large slit would not be a slit at all.

Now, it is time to remember that we let the rate of expansion C in equation (6.2) equal 1. Inserting C back into the phase, we get the following expression for the period T for a linear expansion of the slit:

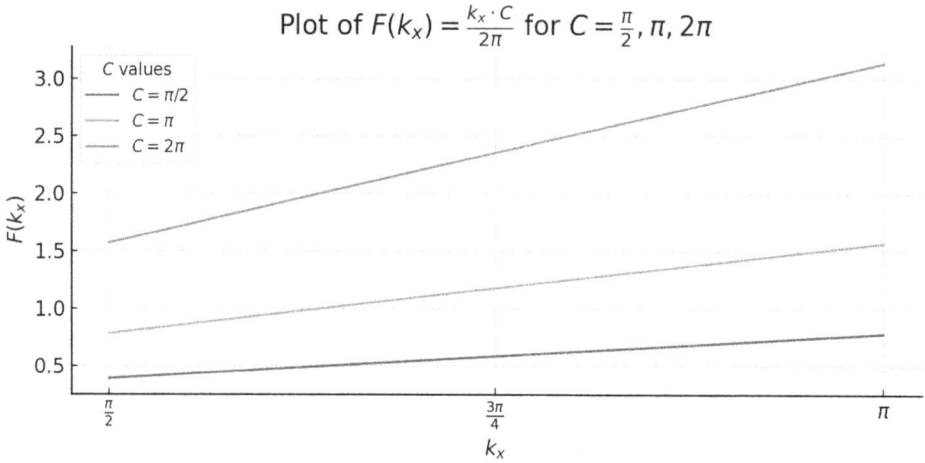

Figure 6.3. Plotting frequency F versus k_x is a linear relationship. Each rate of change C at which the slit expands can be extracted from the slope.

$$T = \frac{2\pi}{k_x \cdot C},\qquad(6.5)$$

Showing that T decreases linearly with the increasing speed C at which the slit widens shows that there is a direct relationship between the proportionality of the motion in real space, where the aperture function exists, and the Fourier transformed k-space, where the diffraction pattern exists. The plot of $F(k_x)$ versus C demonstrates a linear relationship in figure 6.3. Notice that we define C as the rate of one side of the slit moving. Of course, the width of the slit expands at a rate of $2C$ as each side moves in the opposite direction.

6.2.2 Practice problem

Consider an experiment with a slit that expands at a constant rate C. Estimate the expansion rate of the single slit by plotting the frequency F versus k_x using figure 6.4.

6.2.3 Nonlinear changes

Returning to equation (6.7), we reconsider a nonlinear expression for $a(t)$. From the earlier discussion, we established that by examining the distance between the peaks of $I(t)$ at a specific point in the diffraction pattern, for a given k_x, we can infer the rate at which a single slit either expands or contracts. The rate C stays constant for a particular k_x as long as the period T is constant. However, if the period varies for a given k_x, the rate becomes nonlinear. For instance, if $a(t)$ is a quadratic function, $a(t) = Ct^2$, then the period T would increase quadratically over time. In this section, we will explore oscillatory behavior, which frequently appears in natural phenomena, like the movement of a microscopic worm. Consequently, we utilize a sine function to model the oscillation of the slit width, expressed as

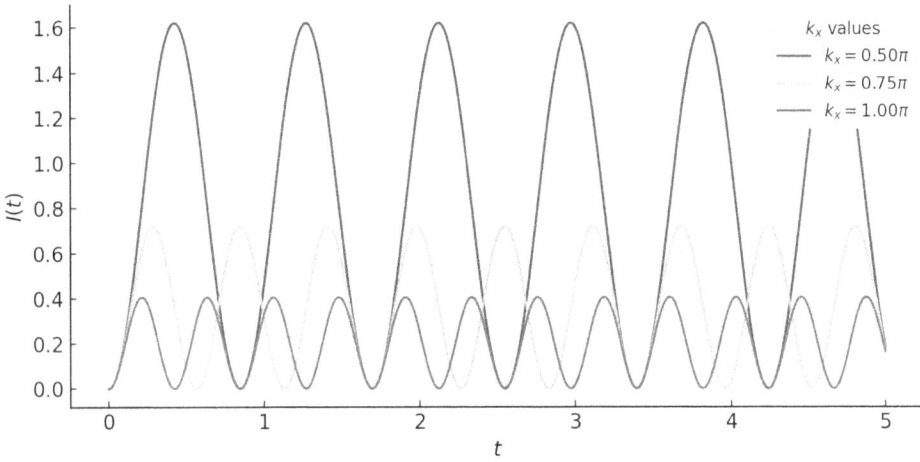

Figure 6.4. Simulated data set at three different points in the diffraction pattern.

$$a(t) = a_0 \sin \omega t, \tag{6.6}$$

where ω represents the angular frequency of the oscillation and a_0 is the amplitude of the oscillation. The modified version of equation (6.7) now reads:

$$I(k_x, t) = E_0^2 a_0^2 \sin^2(\omega t) \, \mathrm{sinc}^2\!\left(\frac{k_x a_0 \sin \omega t}{2}\right). \tag{6.7}$$

Similar to the linear motion illustrated in figure 6.1, the sinc function grows inversely with the widening of the slit; however, the sinc function broadens and the amplitude lowers as the slit narrows again, as shown in figure 6.5.

Again, how does the intensity fluctuation $I(t)$ change at one point in the diffraction pattern. How does $I(t)$ correspond to the motion in real space; i.e., the oscillating slit width? Substituting equation (6.6) into equation (6.4) we get:

$$I(t) = \frac{4}{k_x^2} \sin^2\!\left(\frac{k_x a_0 \sin \omega t}{2}\right). \tag{6.8}$$

Figure 6.6 shows the effect of the oscillatory behavior on $I(t)$. Unlike linear motion, the period is constant. k_x picks up another harmonic for each multiple of π; however, the fundamental (lowest) frequency is always present and matches the oscillatory frequency. The phase in equation (6.8) $k_x a_0 \sin \omega t / 2$ equals zero when k_x is an integer multiple of π. The phase always equals zero when $\sin \omega t$ equals zero, so that the angular frequency ω superimposes its frequency regardless of the values of k_x.

6.3 Useful theorems and lemmas

DOD relies on the relative motion between elements of a diffracting object (aperture function). A common structure is a double slit. Imagine segments of a structure

$$3\text{D plot of } I(k_x, t) = \sin^2(2\pi t)\,\text{sinc}^2\!\left(\frac{k_x \sin 2\pi t}{2}\right)$$

$$I(k_x, t) \text{ for different values of } t$$

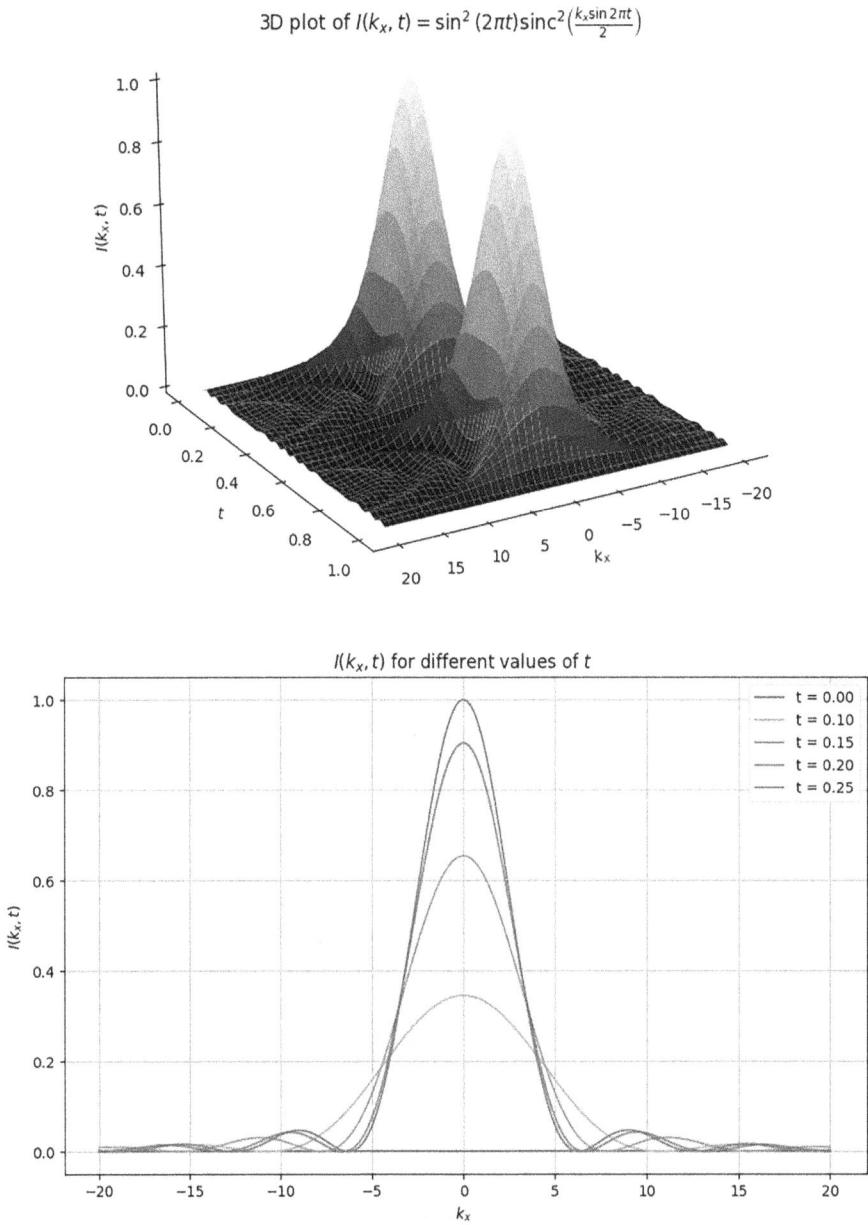

Figure 6.5. Setting the constants E_0 and a_0 equal to 1, the diffraction pattern shows an oscillating interference structure, reflecting the sinusoidal motion in real space also called object space where the aperture function exists. (Top) 3D representation of $I(k_x, t)$ The diffraction pattern due to a single slit with an oscillating slit width $a(t) = a_0 \sin \omega t$. (Bottom) A cross-section for various times between 0 and 1/4 of the cycle as the sinc function grows.

$$I(t) = \frac{4}{k_x^2} \sin^2\left(\frac{k_x \sin 2\pi t}{2}\right)$$

$$I(t) = \frac{4}{k_x^2} \sin^2\left(\frac{k_x \sin \pi t}{2}\right)$$

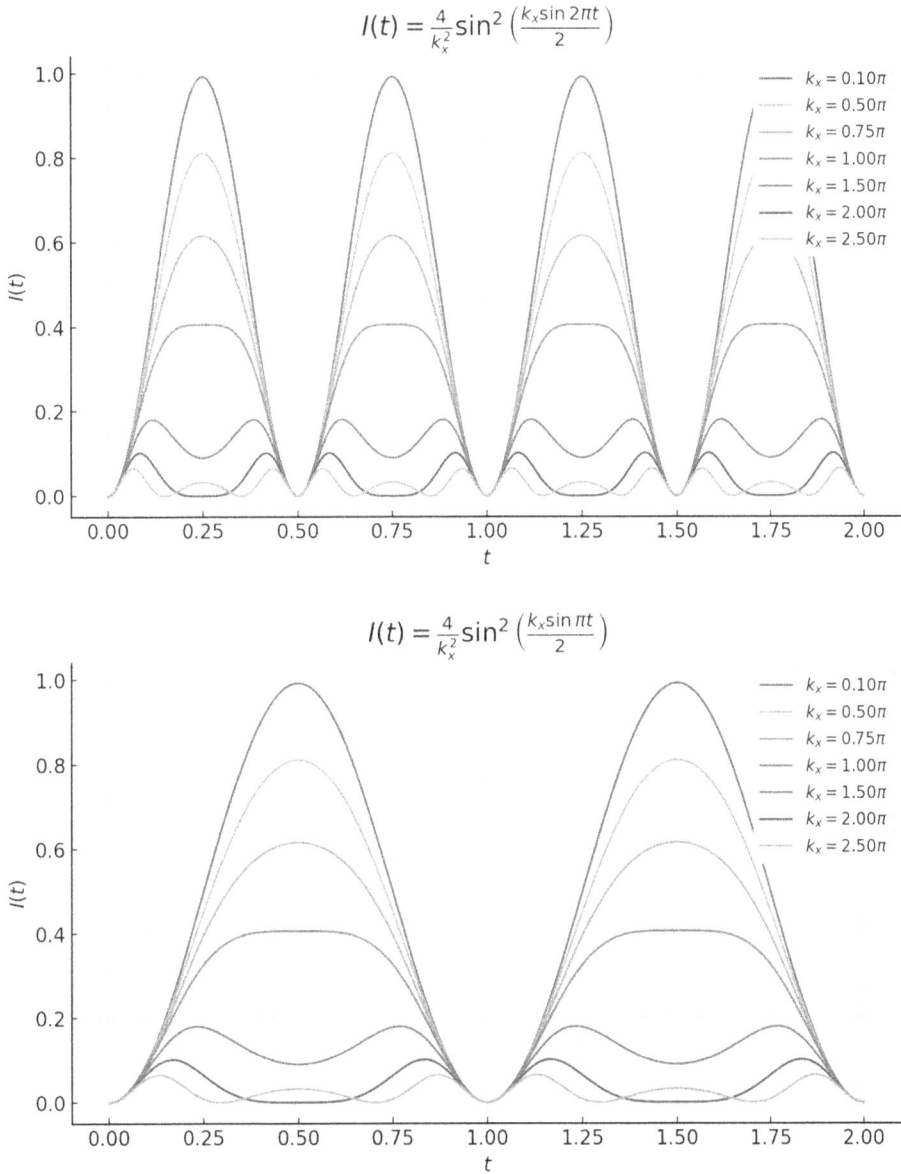

Figure 6.6. Simulated data set at different points in the diffraction pattern for an oscillating single slit for $E_0 = 1$. Additional harmonics are set in for $k_x = n\pi$, where n is an integer. (Top) $\omega = 2\pi$ so that the frequency equals 1. (Bottom) $\omega = \pi$ so that the frequency equals 0.5.

oscillating relative to each other, such as the tail and head of a worm. For this purpose, we first explore some static theorems that are often used to simplify the calculations of diffraction patterns.

6.3.1 Addition theorem

In the practice problem in section 4.3.2, we already employed the addition theorem [1]. Given two aperture functions $g_1(x)$ and $g_2(x)$ with the corresponding Fourier transforms (FTs) $E_1(k)$ and $E_2(k)$, we can calculate the FT $E(k)$ of the sum of $g_1(x)$ and $g_2(x)$, $g(x)$, by adding the FTs $E_1(k)$ and $E_2(k)$ since the integral of a sum is equal to the sum of the integrals of the summands:

$$g_1(x) + g_2(x) \Leftrightarrow E_1(k) + E_2(k), \tag{6.9}$$

where x and k are conjugate variables. This theorem makes it possible to build a modular FT from known FTs.

6.3.2 Shift theorem

We also employed the shift theorem in the practice problem in section 4.3.2 [1] where we used two single slits and each shifted them by a distance b in opposite directions:

$$\begin{aligned} g_1(x + b) &\Leftrightarrow E_1(k)\mathrm{e}^{ikb} \\ g_2(x - b) &\Leftrightarrow E_2(k)\mathrm{e}^{-ikb} \end{aligned} \tag{6.10}$$

Combining the two expressions in equation (6.10), setting $g_1(x + b) + g_1(x - b)$ equal to $g(x)$, and $E_1 = E_2 = E$ we can center symmetrical shifts such as for the double slit so that the FT becomes a real expression:

$$\begin{aligned} g_1(x + b) + g_1(x - b) &\Leftrightarrow E(k)\mathrm{e}^{ikb} + E(k)\mathrm{e}^{-ikb} \\ &\Leftrightarrow E(k)(\mathrm{e}^{ikb} + \mathrm{e}^{-ikb}) \\ &\Leftrightarrow E(k)(\cos(kb) + i\sin(kb) + \cos(-kb) - i\sin(ka)) \\ g(x) &\Leftrightarrow 2E(k)\cos(kb). \end{aligned} \tag{6.11}$$

Notice that $\cos(kb) = \cos(-kb)$ and $\sin(-kb) = -\sin(kb)$ so that the sine adds to zero and the cosine doubles.

A time dependency of b such that $b = b(t)$ results in temporal oscillations regardless of the value of E. The correlation in equation (6.11) shows that the whole expression equals zero whenever the cosine equals zero, thus inducing an oscillation that is directly related to the aperture function.

6.3.3 Example: variable double slit

The practice problem in section 4.3.2, as shown in figure 6.7, is an example of an application to the additions and shift theorem.

The resulting diffraction pattern as established in the practice problem in section 4.3.2 is:

$$I = |E|^2 = 4E_0^2 a^2 \cos^2(k_x b/2)\,\mathrm{sinc}^2(k_x a/2). \tag{6.12}$$

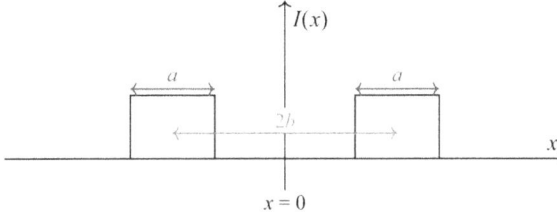

Figure 6.7. Two identical single slits of width a modeled as top-hat functions combine to a double slit with a center-to-center slit separation $2b$. Each slit was shifted by a distance b in opposite directions.

6.4 Dynamic double slit

Now that we have an expression for an oversampled double slit, we can vary the distance b between slits so that the slit separation is a function of time: $b = b(t)$. Equation (6.12) becomes:

$$I(k_x, t) = 4E_0^2 a^2 \cos^2(k_x b(t)/2)\,\mathrm{sinc}^2(k_x a/2). \tag{6.13}$$

Let $4E_0^2 a^2 = I_0$ since it is constant in time and will make the time-dependent calculations clearer:

$$I(k_x, t) = I_0 \cos^2(k_x b(t)/2)\,\mathrm{sinc}^2(k_x a/2). \tag{6.14}$$

There are many ways in which the slit separation might change. The separation between slits might increase at a linear rate, an accelerating rate, exponentially, or even oscillate with some periodicity. In the following, we explore the effects of slit separation to understand how the changes in object space (the space in which the aperture function exists) affect k-space (the Fourier transform).

6.4.1 Linear changes

Similar to the single slit, let us first explore linear changes in the slit separation b of the double slit; i.e., let

$$b(t) = C \cdot t, \tag{6.15}$$

where C is a constant. Equation (6.19) then becomes:

$$I(k_x, t) = I_0 \cos^2\left(\frac{k_x C}{2}t\right)\mathrm{sinc}^2(k_x a/2). \tag{6.16}$$

Figure 6.8 illustrates the progression of an overlapping double slit, $b(t = 0) = 0$, which is just a single slit to a double slit pattern. The sinc function has zero value when the phase $k_x a/2$ carries integer multiples n of π so that:

$$k_x = 2n\pi/a. \tag{6.17}$$

The zeros in figure 6.8 are stationary, since the width of the slits does not change. The peaks inside the single-slit envelope; i.e., the sinc function, is indicative of the

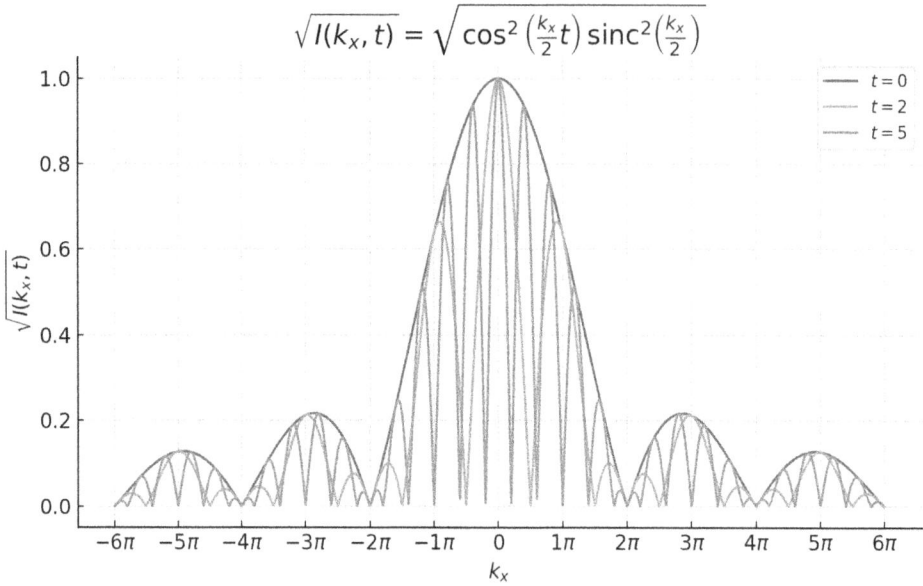

$$\sqrt{I(k_x, t)} = \sqrt{\cos^2\left(\tfrac{k_x}{2}t\right)\operatorname{sinc}^2\left(\tfrac{k_x}{2}\right)}$$

Figure 6.8. Diffraction pattern due to a double slit with width $I_0 = 1$, $a = 1$, and rate at which the slits separate $C = 1$. $t = 0$ represents an overlap of the slits; i.e., a single slit, transitioning into a double slit. Zeros are stationary at a multiple of 2π.

separation of the slits. This motion can be analyzed in the same manner as the linear motion in section 6.2.1. Repeating this procedure, we simulate a PD in the diffraction pattern by plotting the intensity for a particular point in the diffraction pattern fixing k_x. Being careful to pick a point that is not equal to a multiple of 2π, a multiple of π renders a maximum value for the sinc function. Figure 6.9 displays the intensity $I(t)$ for arbitrary points avoiding even integer multiples of π.

6.4.2 Nonlinear changes

Finally, we consider a double slit that oscillates so that the slit width

$$b(t) = b_0 \sin \omega t, \tag{6.18}$$

where b_0 is the oscillation amplitude and ω the oscillation amplitude and equation (6.19) then becomes:

$$I(k_x, t) = I_0 \cos^2\left(\frac{k_x b_0 \sin \omega t}{2}\right)\operatorname{sinc}^2(k_x a/2). \tag{6.19}$$

Similarly to figure 6.5, the diffraction pattern and the frequency change, only this time, within the stationary sinc envelope due to the single slits shown in figure 6.10.

As previously, a PD in the diffraction pattern while avoiding stationary zeros, in this case, even multiples of π, results in a time series of intensity $I(t)$ at one point in the diffraction pattern. The emerging pattern in figure 6.11 is similar to the single slit with an oscillating width.

$$I(k_x, t) = \cos^2\left(\frac{k_x}{2}t\right)\text{sinc}^2\left(\frac{k_x}{2}\right)$$

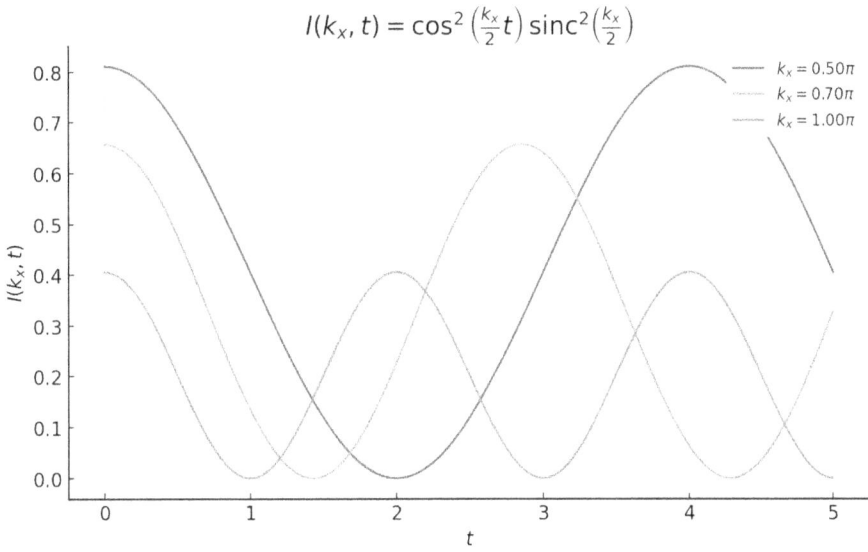

Figure 6.9. Linear double-slit expansion. The slit width increases at a steady rate $C = 1$. Intensity $I(t)$ for select points in the diffraction pattern

$$\sqrt{I(k_x, t)} = \sqrt{\cos^2\left(\frac{k_x\,2\sin(2\pi t)}{2}\right)\text{sinc}^2\left(\frac{k_x}{2}\right)}$$

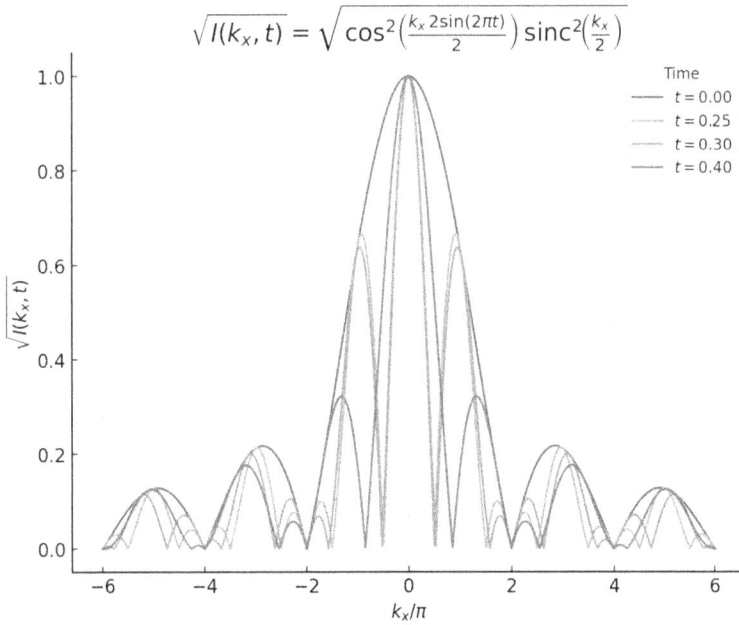

Figure 6.10. Double-slit oscillation. The slit width undergoes simple harmonic motion with amplitude $b_0 = 2$ and the angular frequency $\omega = 2\pi$.

$$\sqrt{I(k_x, t)} = \sqrt{\cos^2\left(\frac{k_x\, 2\sin(2\pi t)}{2}\right) \operatorname{sinc}^2\left(\frac{k_x}{2}\right)}$$

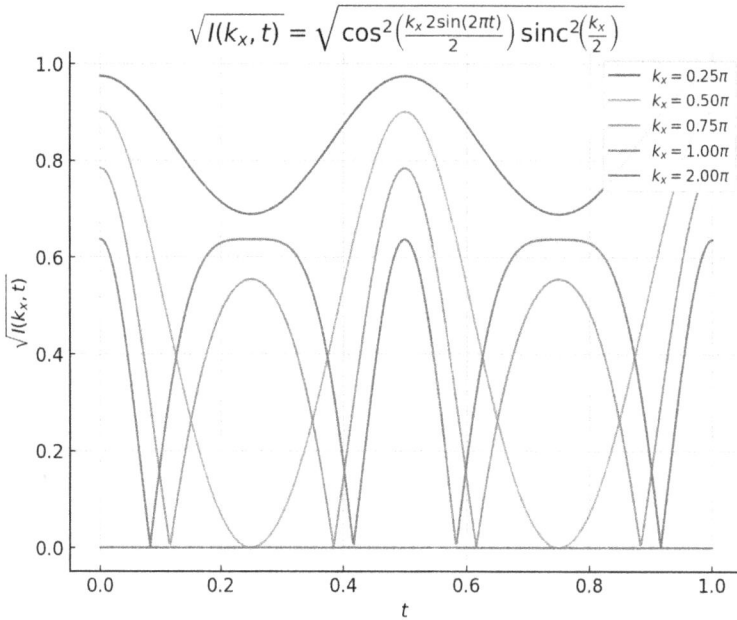

Figure 6.11. Double-slit oscillation. The slit width undergoes simple harmonic motion with amplitude $b_0 = 2$ and the angular frequency $\omega = 2\pi$. Placing a photodiode at a point of stationary destructive interference, in this case $k_x = 2\pi$ renders no signal at all.

Again, the oscillations are consistent with the angular frequency of 2π. The cosine is forced to 1 whenever the sine equals zero, and then the squared cosine doubles the frequency the same way it did in the case of a single slit. The fundamental frequency is 0.5 on the graph, indicating a frequency of 1 in the aperture function (object in real space). The frequency increases as we move farther away from the central maximum; that is, increasing k_x. At certain spatial benchmarks for k_x such as $\pi/2$, we observe frequency doubling.

6.5 Signal analysis using Fourier transforms

The previous sections in this chapter have shown that a time-dependent aperture function $g(x, t)$ is in direct proportion to the diffraction pattern. One might say, well, of course there is a connection since the diffraction pattern is calculated from the aperture function. That statement is true; however, as we saw in the preceding chapters, an FT or discrete FT (DFT) is a complex calculation. The surprising part is the direct proportionality between the two spaces, real space and k-space. We learned how to analyze the time series $I(t)$ because one point in the dynamic diffraction pattern is a superposition of all points of the diffracting object. Of course, for demonstration purposes, we had picked straightforward dynamics such as linear motion and simple harmonic motion.

Signal analysis can become tedious or even impossible for more complex signals. A single point in the diffraction pattern is a superposition of all points in the aperture function so that the time series $I(t)$ reflects all the dynamics of the diffracting object. Imagine many pieces, such as slits of various orientations and dynamics, moving at the same time. In the previous sections, we have been interested in the frequencies represented by the intensity fluctuations of $I(t)$ at one point. An FT is a perfect candidate for this type of analysis, since the FT represents the frequencies and amplitudes of a signal. We discussed FTs, DFTs, and fast FTs (FFTs) in previous chapters. For ease we will now use FFTs to analyze the signals.

Here we consider the cases we examined earlier in this chapter: a linearly expanding single slit, a single slit with an oscillating width, a linearly expanding double slit, and a double slit with an oscillating slit separation.

The magnitude of any FT is a symmetrical structure that stretches from negative to positive numbers (figure 6.12). In this case, we are examining a temporal frequency spectrum of the intensity $I(t)$. We only need the positive values of the

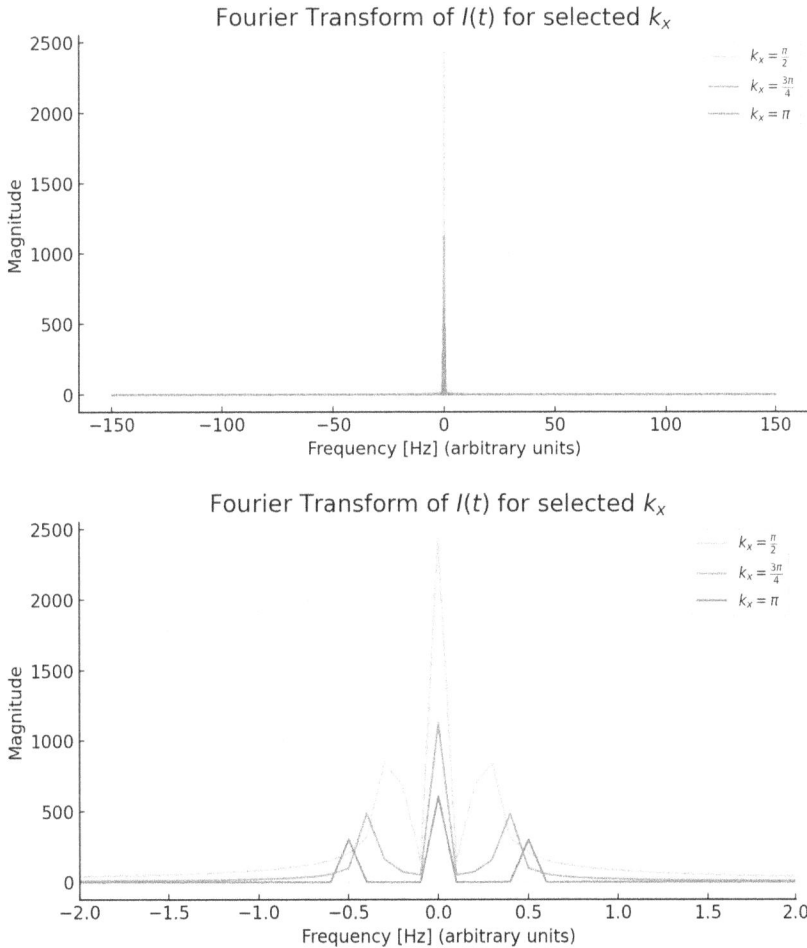

Figure 6.12. FFT of a single slit widening at a constant rate for a total time of $T_0 = 10$ s. (Top) The entire range of the FFT spans 3000 data points. (Bottom) Zoomed in FFT showing a resolution of 0.1 Hz.

FFT. Similarly to examining diffraction patterns, the resolution is critical as well as including critical features. So far, we have dealt with arbitrary units. We will choose units of seconds and Hz in this section to make the physical concepts more relatable.

6.5.1 Resolution

The FFT in figure 6.12 is based on the linear expansion of a single slit represented in figure 6.2. The total data time interval is $T_0 = 10$ s, indicating a frequency resolution $\Delta f = 1/T_0 = 0.1$ Hz. This agrees with the Nyquist theorem, which states that the rate of measurement must be at least twice the rate of the function that is being probed. A longer data acquisition period is necessary for a better resolution, as shown in figure 6.13.

$$I(t) = \frac{4}{k_x^2}\sin^2\left(\frac{k_x t}{2}\right) \text{ for selected } k_x$$

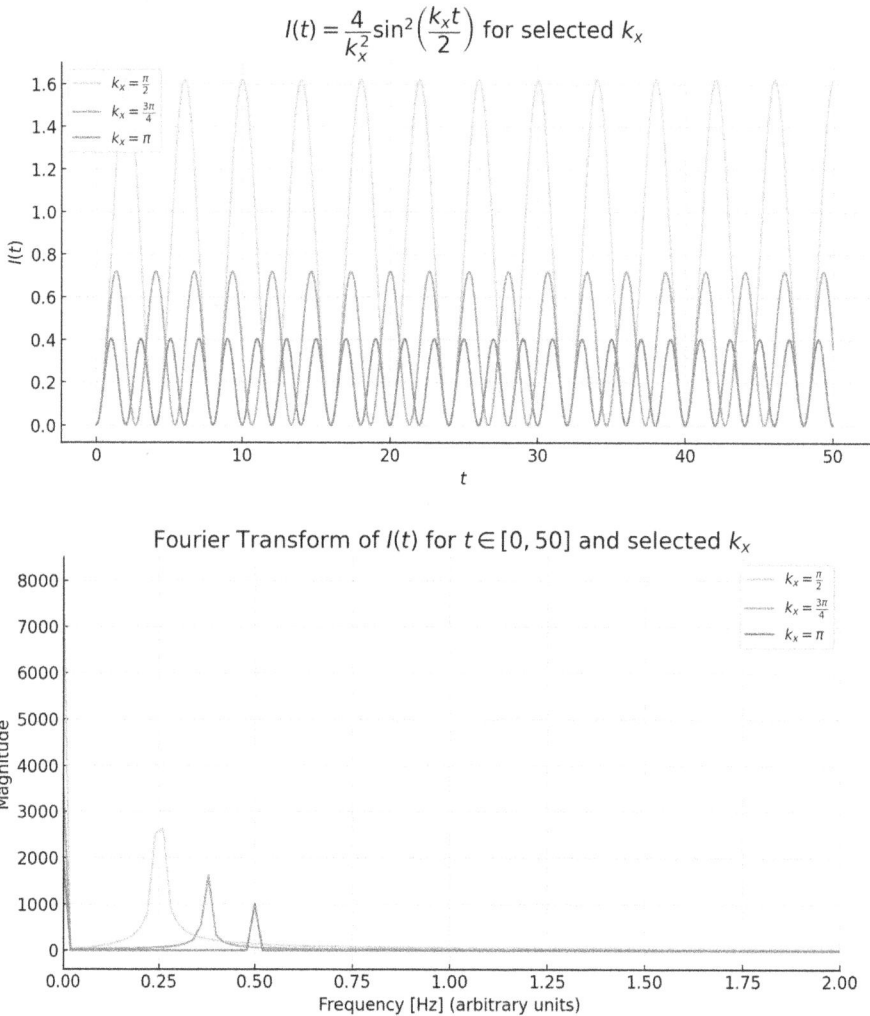

Figure 6.13. FFT of a single slit widening at a constant rate for a total time of $T_0 = 50$ s. (Top) The entire range of the FFT spans 3000 data points. (Bottom) Zoomed in FFT with a resolution of 0.02 Hz.

Even with a high resolution the FFTs associated with each sign function are not delta functions at the expected frequencies. Instead show a rather narrow spike with a finite width because the FFT is a form of a DFT and the discreteness leads to some width in the FFT even though the spikes narrow with increasing resolution.

6.5.2 Frequency range

The range of frequencies covered also has its roots in the Nyquist frequency. Figure 6.13 each curve has 10 000 data points for a period of 50 s. That is a data acquisition rate of 200 Hz; however, Nyquist's theorem states that we need at least twice the frequency to measure an oscillation. That means that effectively, we can only detect frequencies up to 100 Hz. Before deleting the negative frequencies in figure 6.12, we had the same number of data points as in the time series.

6.5.3 Linear motion

We see for all FFTs a spike at zero, indicating that the oscillation has a constant component since the square of any function does not have negative values. This constant value in $I(t)$ does not oscillate and, therefore, has a zero frequency. The time series of the linear motion displays a single frequency at a time without harmonics.

6.5.3.1 Single slit
We already plotted the FFT of a linearly expanding single slit in figure 6.13. The FFT agrees with the time series $I(t)$ for various k_x. Only two frequencies are necessary to determine the speed of linear motion.

6.5.3.2 Double slit
The time series $I(t)$ in figure 6.9 of a linearly expanding double slit; i.e., the slit separation expands linearly. In this example, we took 50 s of the time series to ensure proper resolution and only plotted the relevant section of the FFT in figure 6.14.

In the case of a single linear change in the diffraction pattern, a Fourier transform may not always be necessary; however, imagine a two-dimensional space, for example, with more than one linear motion, and in that case, an FFT will be very helpful in sorting out frequencies. Each FFT has only one peak, since each location of k_x is associated with a single frequency for linear motion because the location sets the frequency (see equation (6.19)).

6.5.4 Nonlinear motion

Now we revisit the single slit with an oscillating width and the double slit with its oscillating slit separation.

6.5.4.1 Single slit
For the single slit we use $\omega = 2\pi$ so that the frequency equals 1 Hz, as shown at the top of figure 6.6. In the time series that frequency doubles since the sine is squared.

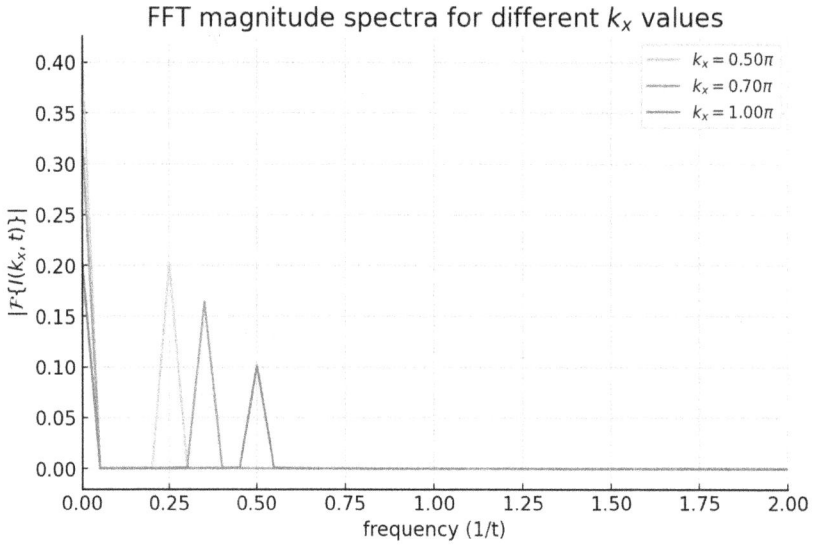

Figure 6.14. Double-slit expansion. The slit separation increases at a linear rate. The peaks are the frequencies at various positions k_x.

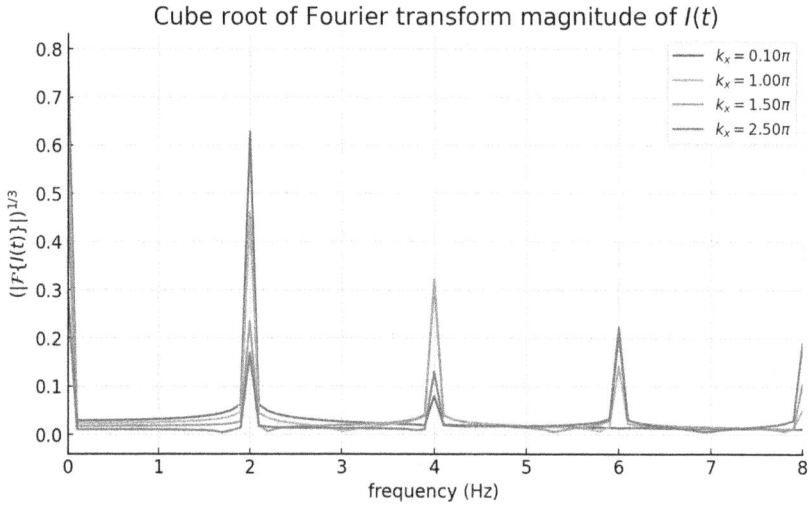

Figure 6.15. FFT of a time series $I(t)$ for a single slit oscillation of 1 Hz. Each time series has several harmonics for various positions of k_x.

This is reflected in the FFT in figure 6.15. The FFT now has several peaks for each position k_x. The position closest to the central maximum produces the largest peak for the fundamental frequency with diminishing harmonics. The dominant frequency increases with k_x.

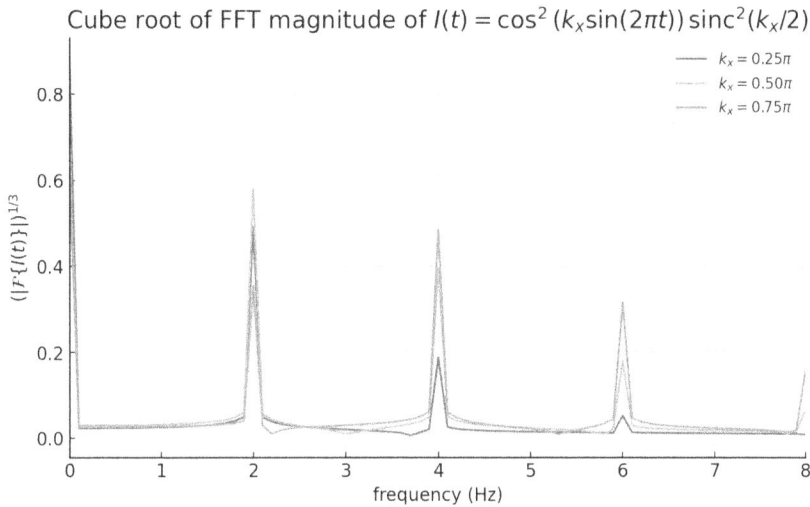

Figure 6.16. FFT of a time series $I(t)$ for a double slit with and oscillating slit separation of 1 Hz. Each time series has several harmonics for various positions of k_x.

6.5.4.2 Double slit

Similarly, for the single slit, there are multiple peaks for each time series $I(t)$ for oscillating conditions. The oscillating double slit is a little more complex as it starts as a single slit that separates into a double slit. However, in general, the fundamental frequency dominates closer k_x to the central maximum in figure 6.16.

In general, the hallmark of linear motion is FFT with a single peak for each position k_x, while simple harmonic motion displays several peaks for each position k_x with the strongest fundamental frequency peak closest to the central maximum. Different nonlinearities are linked to different FFT patterns.

6.5.5 Practice problem

Consider a single slit that expands at an accelerating rate so that the width of the slit is $a(t) = Ct^2$, where $C = 1$. Establish an equation for the diffraction intensity $I(k_x, t)$. Plot $I(k_x)$ for at least three different time values t. Plot $I(t)$ for at least three different position values of k_x. Plot the FFT for all three values of k_x. What is the hallmark of a quadratic expansion? How can you calculate the expansion constant C?

Reference

[1] James J F 2011 *A Studentas Guide to Fourier Transforms: With Applications in Physics and Engineering* (Cambridge: Cambridge University Press)

IOP Publishing

Optical Interference and Dynamic Diffraction
Research methods for undergraduates
Jenny Magnes and Juan M Merlo-Ramírez

Chapter 7

Application of dynamic optical diffraction

7.1 Introduction

Dynamic optical diffraction (DOD) is an emerging technique that leverages coherent light scattering to probe the locomotion and structural dynamics of microscopic organisms. Unlike conventional microscopy, DOD does not rely on imaging planes or high magnification optics. Instead, it translates organismal motion into a time-dependent diffraction pattern that can be sampled at single or multiple points to yield a one-dimensional time series containing the full dynamics of the system. This approach enables the analysis of movement in three dimensions, free from the focal constraints of traditional microscopy, while simultaneously revealing nonlinear and chaotic features embedded in biological locomotion [1–4]. DOD lends itself to nonlinear studies of microscopic systems. We use a microscopic worm called *C. elegans* to demonstrate applications in nonlinear dynamics without detailing the field of nonlinear dynamics [5, 6].

In this book, we focus on far-field (Fraunhofer) diffraction. There are two basic types of optical far-field diffraction:

- diffraction by undersampling;
- diffraction by oversampling.

Diffraction by undersampling is used in crystal structures. In this case, the diffracting optical beam is smaller than the sample and the oversampling ratio σ (equation (5.28)) can be less than 2. The crystal structure provides small scatterers at regular intervals within the sampling area that provide enough signal (figure 7.1). The decoding of the diffraction signal is often completed using an Ewald sphere [7] that maps the diffraction pattern in a reciprocal space linked to a Fourier space.

Diffraction by oversampling tends to be used in quality control, vibration monotoring, and image reconstruction using diffraction patterns. In this case, the diffracting optical beam is larger than the sample and the oversampling ratio σ

doi:10.1088/978-0-7503-4836-2ch7

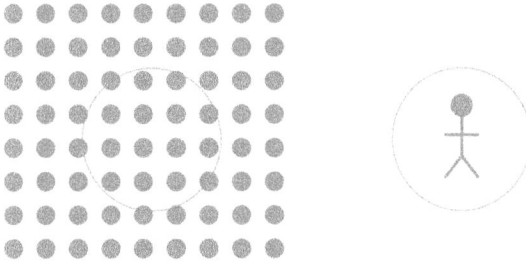

Figure 7.1. Two diffraction methods: (left) the periodic structure only partially covered by the laser shown in pink is undersampled as the dots are the diffracting object and typically take up more area than the laser light. (right) The structure is oversampled by the laser beam fully enveloping the structure and occupying less than half the area of the laser beam.

(equation (5.28)) must be more than 2 (figure 7.1). This text focuses on oversampling techniques, as they are the most explored in the application of DOD.

7.2 Principles of dynamic optical diffraction

At the core of DOD is the Fraunhofer diffraction principle, where the intensity distribution in the far-field is proportional to the squared modulus of the Fourier transform of the diffracting object. This method is easily applied to live microscopic species as the diffraction effects are most pronounced in the visible light range when dealing with objects on the micrometer scale. Monochromatic coherent light is very accessible in the visible range through lasers. Even laser pointers provide enough power and coherent properties for this type of diffraction. However, DOD works at any wavelength as long as the size of the oversampled object matches the wavelengths used. For example, microwaves could be used on an object the size of a coffee cup.

In this chapter, we use a live nematode such as *C. elegans* to illustrate applications; nevertheless, this is only one possible use of DOD. A live nematode such as *C. elegans*, changes body shape and oscillatory motion modulating the diffraction pattern in real time. Recording intensity fluctuations at a fixed off-axis point in the diffraction field yields a time series reflecting the superposition of light scattered from all points along the organism (figure 7.2).

Mathematically, the diffraction signal at a point $\overrightarrow{k} = (k_x, k_y)$ in Fourier space can be expressed using a more general version of the discrete Fourier transform (DFT) equation (5.3) accounting for more than one dimension and the temporal dependency:

$$X(t)_{\mathbf{k}} = \sum_{n=0}^{N-1} x(t)_{\mathbf{n}}\, e^{2\pi i \mathbf{k}\cdot\mathbf{n}/N}, \tag{7.1}$$

where $x(t)_n$ represents the optical density in the object space, and $X(t)_{\overrightarrow{k}}$ is the field amplitude in diffraction space. The detected intensity is

C. elegans Optical Diffraction Setup

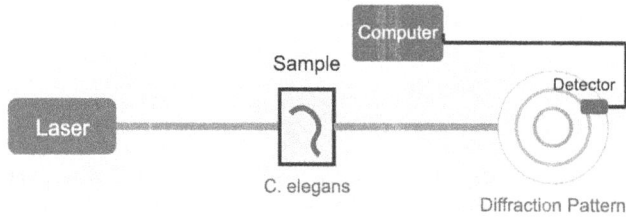

Figure 7.2. Schematic of a DOD setup. A coherent laser beam oversamples a freely swimming *C. elegans*, producing a far-field diffraction pattern. The intensity is recorded either by a photodiode or by a CMOS camera at selected channels.

$$I(t)_{\mathbf{k}} \propto |X(t)_{\mathbf{k}}|^2. \tag{7.2}$$

Thus, temporal changes in the geometry of the worm appear as measurable intensity fluctuations.

7.3 Experimental implementations

Early implementations used single-point detection, where a photodiode sampled the diffraction intensity at a fixed location [1]. This approach demonstrated that locomotory oscillations appear as discrete frequency peaks corresponding to thrashing frequencies. Later refinements introduced multichannel detection using CMOS cameras, enabling simultaneous acquisition across the diffraction pattern and validating that the dynamic parameters are independent of the sampling position [4].

A typical setup, as shown in figure 7.2, includes:

- A low-power He–Ne laser (632.8 nm) with a sufficiently large beam diameter to oversample the worm.
- A quartz cuvette with freely swimming *C. elegans*.
- Detection via photodiode or CMOS array.
- Digital acquisition for time-series and nonlinear analysis.

The wavelength is set not to interfere with biological life. 632.8 nm is red, almost infrared, and is known not to interfere with the locomotion of microorganisms. Of course, there could be reasons to intentionally change the wavelength. For example, if one wants to study the effect of shorter wavelengths on microscopic life.

7.4 Frequency analysis of locomotion

The Fourier spectrum of the diffraction time series reveals the dominant frequencies of microscopic structures as in the example of locomotory frequencies of *C. elegans* in figure 7.3. Resonance peaks correspond to thrashing motions, with higher harmonics emerging as the detectors sample further from the diffraction maximum.

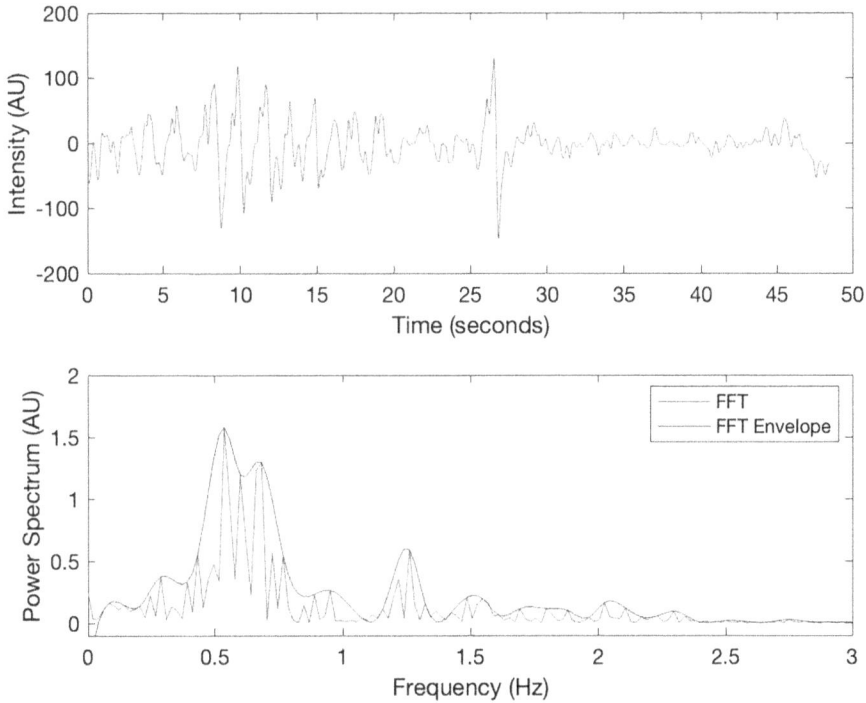

Figure 7.3. A live intensity time series at one point in a diffraction pattern from a thrashing *C. elegans* (top) and its corresponding FFT zoomed in on the largest components (bottom) with fitted envelope to the FFT.

Importantly, DOD detects frequencies at sub-wavelength sensitivity, even when visual microscopy fails to resolve them across multiple length scales. Traditional microscopy gives insight into structural, behavioral, and neural information; however, these observations are one length scale at a time. DOD captures multiple length scales simultaneously so that orchestrated dynamics can be analyzed.

The envelope in the bottom graph of figure 7.3 indicates what the FFT might look like with an infinitely long time series as the frequencies drift, capturing multiple time scales simultaneously. The peaks indicate major and minor thrashing frequencies of the worm in relation to various segments in its body.

7.5 Nonlinear and chaotic markers

Beyond frequency analysis, DOD time series can reveal chaotic features of locomotion. Using tools from nonlinear dynamics such as mutual information (MI) [8], false nearest neighbors (FNN) [9], and Takens embedding theorem [10], researchers reconstruct phase trajectories. Positive values of the largest Lyapunov exponent (LLE) indicate sensitive dependence on initial conditions as phase trajectories diverge exponentially. A positive LLE is a marker fo deterministic chaos [11], with measured values around $1.2 \quad s^{-1}$ for adult wild-type *C. elegans* [3, 4]. DOD signals reveal irregularities not attributable to noise but to deterministic

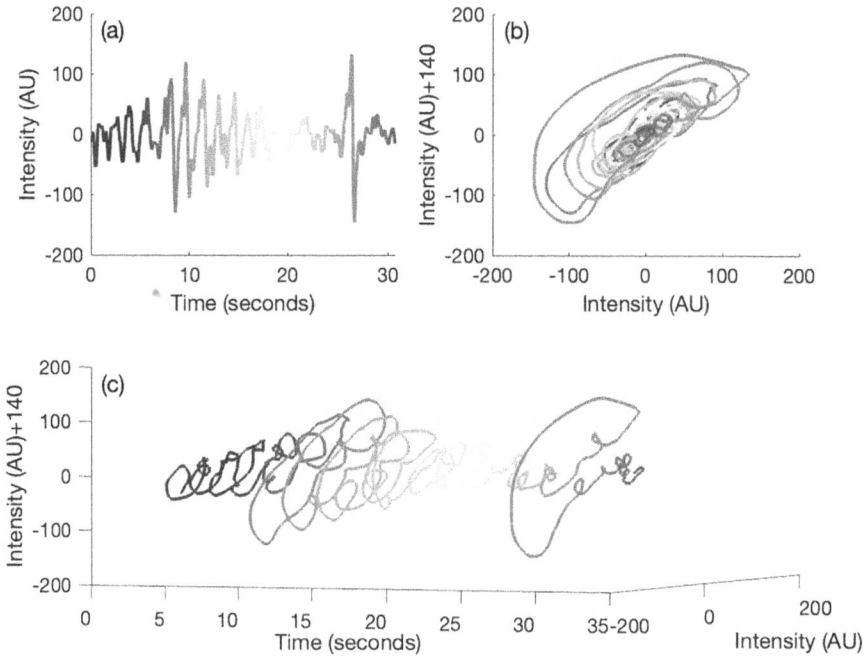

Figure 7.4. DOD diffraction due to a live *C. elegans* with 40 000 data points. (a) Intensity time series due to one point in the diffraction pattern. (b) phase space trajectory using a lag of 140 data points projected onto a two dimensional space. (c) Three-dimensional representation combining time and phase space representation showing the evolution of the attractor.

chaos. Using Takens' embedding theorem, a time series $I(t)$ can be embedded in a phase space (figure 7.4):

$$\mathbf{X}(t) = (I(t), I(t + \tau), I(t + 2\tau), ..., I(t + (m - 1)\tau)), \qquad (7.3)$$

where τ is the delay and m is the embedding dimension. The LLE quantifies chaos:

$$\lambda_L = \lim_{t\to\infty} \frac{1}{t} \ln \frac{\delta D(t)}{\delta D_0}, \qquad (7.4)$$

where δD_0 is the initial separation of the trajectories in phase space and $\delta D(t)$ their divergence at time t. A positive λ_L indicates sensitivity to initial conditions, the hallmark of chaos. Figure 7.4 shows the (a) intensity time series of a dynamic diffraction pattern at a point in the diffraction pattern and the (b) reconstructed phase trajectories.

Chaotic markers include:
1. Broadband frequency spectra (aperiodicity).
2. Divergence of trajectories in phase space (positive LLE).
3. Independent channels of dynamic parameters, ensuring robustness.

7.6 Advantages over traditional microscopy

DOD offers several advantages:
1. Three-dimensional locomotion tracking without confocal constraints.
2. Real-time dynamics at sub-wavelength sensitivity.
3. Compact, low-cost setup compared to high-speed video microscopy.
4. Access to nonlinear descriptors, such as Lyapunov exponents, typically inaccessible in standard imaging.

7.7 Broader applications

While most work has focused on *C. elegans*, DOD principles extend to:
- Other microscopic organisms (sperm, protists).
- Phenotypic screening for drug or genetic effects.
- Neuroscience modeling of locomotion as emerging from neuronal networks.
- Materials science, for studying nanoscale periodic deformations.
- Materials science for studying crystal growth.

7.7.1 Practice problem

A diffraction time series yields an FFT peak at 1.2 Hz and a secondary peak at 2.4 Hz. Interpret these results in terms of worm locomotion. How might the presence of additional peaks at 0.6 Hz and 1.8 Hz be explained?

7.8 Conclusion

DOD represents a transformative approach to studying microscopic locomotion. By condensing complex three-dimensional motion into analyzable one-dimensional time series, DOD not only complements traditional microscopy but also opens pathways into nonlinear and chaotic analysis of living systems. Its robustness across channels and applicability to multiple scales make DOD a powerful platform for exploring complexity in biology and beyond.

References

[1] Magnes J, Hastings H and Raley-Susman K 2012 Quantitative locomotion study of freely swimming micro-organisms using optical diffraction Open *J. Biophys.* **2** 91–6
[2] Magnes J, Hastings H and Raley-Susman K 2018 Dynamic diffraction analysis of locomotion in C. elegans Open *J. Biophys.* **8** 103–16
[3] Magnes J, Hastings H, Hulsey-Vincent M, Congo C, Raley-Susman K, Singhvi A, Hatch T and Szwed E 2020 Chaotic markers in dynamic diffraction *Appl. Opt.* **59** 6642–8
[4] Zanetti R F, Canavan K L, Zhang S G and Magnes J 2023 Multi-channel measurements of C. elegans' largest Lyapunov exponents using optical diffraction *Appl. Opt.* **62** 6350–64
[5] Strogatz S H 2001 *Nonlinear Dynamics and Chaos: with Applications to Physics, Biology, Chemistry, and Engineering (Studies in Nonlinearity)* **Vol 1** (Boulder, CO: Westview Press)
[6] Feldman D P 2012 *Chaos and Fractals: An Elementary Introduction* (Oxford: Oxford University Press)

[7] Barbour L J 2018 Ewaldsphere: an interactive approach to teaching the ewald sphere construction *Appl. Crystallogr.* **51** 1734–8

[8] Fraser A M and Swinney H L 1986 Independent coordinates for strange attractors from mutual information *Phys. Rev.* A **33** 1134–40

[9] Abarbanel H D I and Kennel M B 1993 Local false nearest neighbors and dynamical dimensions from observed chaotic data *Phys. Rev.* E **47** 3057–68

[10] Takens F 1981 Detecting strange attractors in turbulence *Dynamical Systems and Turbulence* (Lecture Notes in Mathematics) (vol 898) ed D Rand and L-S Young (Berlin: Springer) 366–81

[11] Skokos C 2016 Lyapunov analysis: from dynamical systems theory to applications *Chaos Detection and Predictability* Springer Lecture Notes in Physics vol 916 (Berlin: Springer)

IOP Publishing

Optical Interference and Dynamic Diffraction
Research methods for undergraduates
Jenny Magnes and Juan M Merlo-Ramírez

Appendix A

Optical sources

A.1 Introduction

In this chapter, we will explore the different kinds of light sources, in particular, those that are used for undergraduate research in diffraction, i.e. lasers and thermal sources, though a general overview of sources in the whole electromagnetic spectrum will be provided. Important concepts will be also reviewed, among them, optical power, intensity, polarization, and coherence. A general overview is provided of the generation of electromagnetic waves.

A.2 Useful definitions

A.2.1 Power

As mentioned in section 2.8, light is an electromagnetic field, meaning the energy carried out can be expressed as $\varepsilon_E = 1/2\varepsilon_0 \left|\overrightarrow{E}\right|^2$ in terms of the electric field $\overrightarrow{E}$, or $\varepsilon_B = 1/2\mu_0 \left|\overrightarrow{B}\right|^2$ in terms of the magnetic field $\overrightarrow{B}$, when the wave is moving through free space.

Optical power (P), also known as power in an optics environment, refers to the amount of radiant energy carried by an optical signal per unit time. The accepted unit for optical power is the watt (W). It is possible to determine the power by calculating the ratio between the energy, emitted or received in a certain place, by unit time as [1]:

$$P = \varepsilon/t \tag{A.1}$$

with, ε the energy and t the time.

A.2.2 Intensity

Optical intensity, also known as intensity (I) in an optics environment, provides a measure of the optical power per unit area emitted or received by a surface. The

doi:10.1088/978-0-7503-4836-2ch8
A-1

intensity can be defined as the ratio between the power, emitted or received, and the area (A), i.e.,

$$I = \frac{P}{A}.$$
(A.2)

It is also possible to determine the intensity as defined by equation (2.42) in section 2.7. In this case, as the expression of the field is known, the intensity is calculated as:

$$I(\mathbf{r}) = \mathbf{E}(\mathbf{r})\mathbf{E}^*(\mathbf{r}).$$
(A.3)

In the case of an electromagnetic wave propagating in vacuum, the intensity can be expressed as [1]:

$$I = \frac{1}{2}c\varepsilon_0 \left\langle \vec{E} \right\rangle^2,$$
(A.4)

with c the speed of light in vacuum, ε_0 the permittivity of free space, and $\langle\rangle$ the time average.

A.2.2.1 Intensity from a point source

In the case of a point source emitting light isotropically, wavefronts have spherical shape; the intensity at a distance r from the source can be calculated as:

$$I = \frac{P}{4\pi r^2}$$
(A.5)

A.2.3 Polarization

Polarization refers to the direction in which the electric field oscillates in an electromagnetic field. There are several kinds of polarization: random (unpolarized), linear, and elliptical [1].

In *randomly* polarized light, the electric field that has no preferential direction of oscillation. On the other hand, in *linearly* polarized light, the electric field oscillates in a fixed plane. Finally, in *elliptically* polarized light, the electric field rotates describing an ellipse. In this case, it is said that the polarization is *circular* when the amplitude of the electric field is isotropic along one rotation. It is clear that in such a case, we could have *right* and *left* polarizations. Figure A.1 shows the aforementioned polarizations.

A.2.4 Angular divergence

Angular divergence, also known as beam divergence, is a measure of the change of the radius of a light beam along its propagation. Most of the optics applications require small angular divergences for a better control and confinement of the beams. The angular divergence can be calculated as:

$$\Theta = \arctan\left(\frac{R_f - R_i}{2l}\right),$$
(A.6)

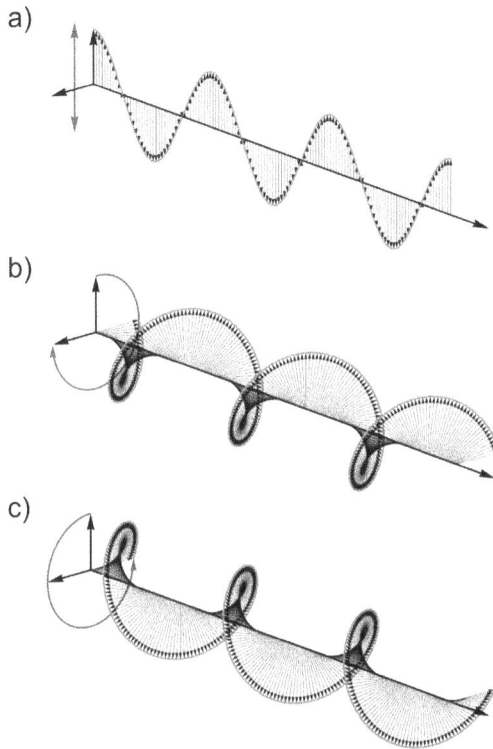

Figure A.1. Different types of polarization. (a) Linear polarization. (b) Right circular polarization. (c) Left circular polarization. In all the cases, the blue arrow represents the averaged electric field direction.

where R_f and R_i are the final and initial beams, respectively, and l the propagation distance.

A.2.5 Emission spectrum

The emission spectrum of a light source refers to the light frequencies emitted by the source as the result of electronic transitions. Among the common sources used in optics experiments, we can find the following emission spectra:

- Black-body. Usually associated with hot bodies emitting a broad spectrum, see figure A.2(a).
- Fluorescence. Associated with molecular transitions. This emission needs a primary source, not necessarily heat, to generate the emission, see figure A.2(b).
- Atomic emission. Associated to the electronic transitions in monoatomic or diatomic molecules of elements, see figure A.2(c).
- Laser emission. Associated to specific electronic transitions under a stimulated radiation process, see figure A.2. This emission is commonly considered monochromatic.

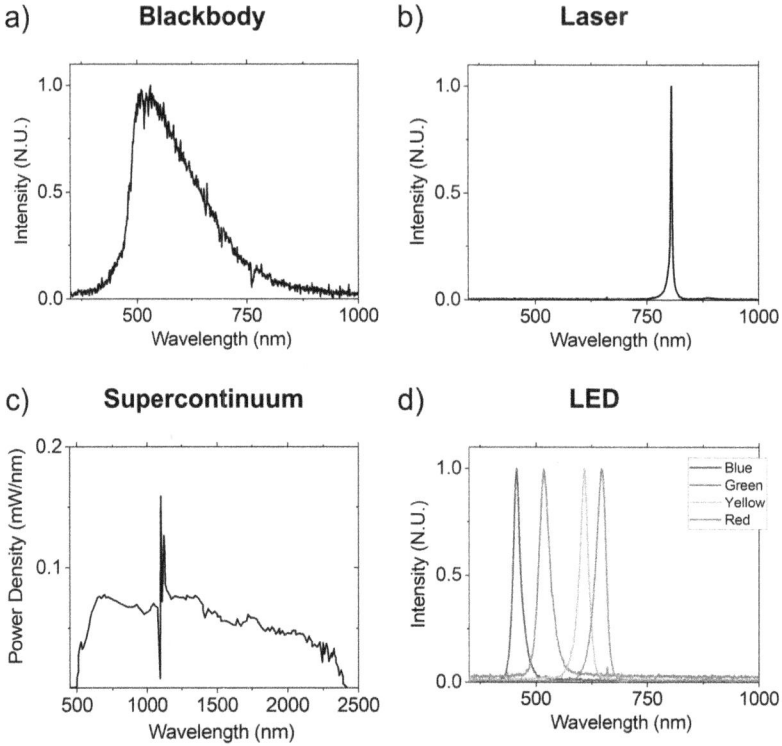

Figure A.2. Intensity spectra of different sources. (a) Blackbody, (b) Laser with wavelength of 810 nm, (c) Supercontinuum, and (d) LED of different collors.

A.2.6 Coherence

The coherence of an optical source refers to the degree of correlation between the phases of different electromagnetic waves emitted by the source [2]. A perfectly coherent source emits waves with a constant phase difference between them, leading to well-defined interference patterns. In contrast, a partially coherent source emits waves with random phase differences, resulting in less distinct interference patterns. The coherence of a source has a quantum origin, however, there is a way to measure it with classical optics, i.e., by knowing the visibility of an interference pattern generated by the source under study. The visibility is calculated as follows:

$$V = \frac{I_{\max} - I_{\min}}{I_{\max} + I_{\min}}$$ (A.7)

where $I_{\max}$ and $I_{\min}$ are the maximum and minimum intensities of the interference pattern, respectively.

A.3 Thermal sources

Thermal sources generate light in a broad wavelength spectrum following the blackbody radiation. This means that the most important factor for the spectrum is the temperature of the source. The usual thermal sources are usually incandescent light bulbs made of different elements. Figure A.2(a) shows the emission spectrum of the Sun (~5700 K). Due to their thermal nature, thermal sources usually emit incoherent and unpolarized light[1].

A.4 Laser sources

The laser (Light Amplification by Stimulated Emission of Radiation) light is generated by the excitation of a material, *gain medium*, by an external source. This produces excited states in the gain medium generating *population inversion*. When these excited states return to their ground state, there is emission of photons that can trigger further excitation of states, and the process repeats; the new photons emitted by this process have the same frequency and phase than the original ones, i.e., highly coherent light. In order to increase the chances for this process to exist, a pair of semitransparent mirrors are added, creating the so-called *laser cavity* [1].

Due to its excitation process, laser light is highly directional, monochromatic, and coherent with a small angular divergence. Figure A.2(b) shows the spectra of a laser diode with wavelength of 810 nm.

Nowadays, lasers are built with different gain media as gases, liquids, semiconductors, and specially doped optical fibers [1]. In recent years, broadband laser sources have become available in the form of supercontinuum lasers, based in nonlinear processes in optical fibers. Figure A.2(c) shows the spectra of this kind of source [3].

A.5 Semiconductor sources

Although the majority of experiments in optics are performed with laser sources, there are some cases in which the high coherence of these sources would become an issue. As such, other light sources based on semiconductor junctions are used. Among these sources, light emitting diodes (LEDs) [4], super-luminescent and diode sources (SLDs). In both cases, the emission spectrum is broad and the coherence length is in the order of microns. Figure A.2(d) shows the spectra of several LEDs.

References

[1] Hecht E 2002 *Optics* (Boston, MA: Addison-Wesley)
[2] Born M and Wolf E 1999 *Principles of Optics: Electromagnetic Theory of Propagation, Interference and Diffraction of Light* 7th edn (Cambridge: Cambridge University Press)
[3] Alfano R R 2022 The Supercontinuum Laser Source *The Ultimate White Light* 4th edn (Berlin: Springer)
[4] Abushagur M A G *Applied Photonics: An Introduction for Physicists and Engineers* (Berlin: Springer)

---

[1] The light emitted by the Earth's atmosphere is polarized due to scattering.

IOP Publishing

Optical Interference and Dynamic Diffraction
Research methods for undergraduates
Jenny Magnes and Juan M Merlo-Ramírez

Appendix B

Photodetectors

B.1 Introduction

In this chapter, we will review the concept of photodetection and the photodetectors that are widely used in the optics research. Although Si and GaAs photodetectors are the most commonly used in the visible and near-infrared spectrum, we will provide examples of photodetectors in the whole electromagnetic spectrum. We present a general overview of detection of electromagnetic waves and semiconductor photodetectors.

B.2 Useful definitions

B.2.1 Spectral response

The spectral response ($R(\lambda)$) of a device or material is a measure of its sensitivity to different wavelengths of light. It is expressed as:

$$R(\lambda) = \frac{S(\lambda)}{E(\lambda)} \tag{B.1}$$

where: $S(\lambda)$ is the signal produced by the device or material at wavelength λ, $E(\lambda)$ is the incident power on the device or material.

B.2.2 Fluence

Fluence (F)) represents the radiant energy per unit area incident on or passing through a surface. The fluence can be expressed as [1]:

$$F = \frac{\mathrm{d}E}{\mathrm{d}A} \tag{B.2}$$

where E the incident energy, and A is the area.

doi:10.1088/978-0-7503-4836-2ch9

B.2.3 Radiance

Optical radiance ($L(\lambda)$) is the power per unit solid angle emitted, reflected, transmitted, or received by a surface per unit projected area and unit wavelength interval. It can be expressed as:

$$L(\lambda) = \frac{d^2\Phi(\lambda)}{dA\ d\Omega\ d\lambda\ \cos(\theta)} \tag{B.3}$$

with Φ is the radiant power, A is the measurement area, Ω is the solid angle, θ is the angle between the normal to the surface and the direction of the radiation.

B.3 Semiconductor detectors

Photodetection on semiconductors is based on the photoelectric effect [2] where a photon with enough energy can kick an electron out of the material and generate an electric current.

Some of the most common uses of photodetectors based on semiconductors are the following [1, 2]:

- Photodiodes:
    Silicon photodiodes are common in visible and near-infrared applications, while other materials like InGaAs are suitable for longer-wavelengths.
- Avalanche photodiodes (APDs):
    Similar to silicon-based APDs, these devices are semiconductor-based and take advantage of the avalanche multiplication effect to achieve higher sensitivity. APDs are used in low-light conditions.
- Phototransistors:
    These are semiconductor devices that combine the properties of a photodiode and a transistor. Phototransistors amplify the photocurrent, providing a larger output signal compared to a standalone photodiode.
- CCD:
    A CCD (charge-coupled device) photodetector is also a semiconductor-based detector. It consists of an array of light-sensitive pixels arranged in rows and columns on a silicon chip. When light strikes the pixels, it generates electric charge proportional to the intensity of the incident light. The operation of a CCD photodetector involves transferring the accumulated charge from each pixel to adjacent pixels in a controlled manner through the process of charge-coupling.
- CMOS:
    A CMOS (complementary metal-oxide-semiconductor) photodetector is also a semiconductor device. A CMOS photodetector is up to 40% more sensitive than CCD detectors.

B.4 Noise

Noise in electronic devices refers to unwanted fluctuations in the output signal. Although the noise has a quantum origin, it is always treated as external interferences. Noise is an important issue in photodetection, particularly at low levels.

The most common type of noise is *thermal noise*, inherent to matter, which arises from the random motion of charge carriers within the photodetector material. It can be characterized by the equation:

$$V_{\text{rms}} = \sqrt{4kTR\Delta f} \tag{B.4}$$

where k is the Boltzmann constant, T the temperature (in K), R the resistance (in Ω), and Δf the frequency bandwidth (in Hz).

References

[1] Hecht E 2002 *Optics* 4th edn (Boston, MA: Addison Wesley)
[2] Abushagur M A G 2025 *Applied Photonics: An Introduction for Physicists and Engineers* (Graduate Texts in Physics) (Berlin: Springer)

IOP Publishing

Optical Interference and Dynamic Diffraction
Research methods for undergraduates
Jenny Magnes and Juan M Merlo-Ramírez

Appendix C

Optical hardware

C.1 Introduction

In this chapter, we will review the most common optical components used in optics research, particularly in the main topic of this book, i.e., dynamic diffraction. Their fundamental principles and common use will also be provided. Among these components, we will study lenses, mirrors, and polarizers, among others.

Here, we show the most commonly used optical elements in dynamic diffraction experiments and research. Among them are lenses, mirrors, polarizers, beam expanders, beam reducers, and spatial filtering.

C.2 Lenses

A lens is an optical device that controls light by converging or diverging light rays using refraction [1]. In practice, a lens can be simplified to a *thin lens*, where it is assumed that light rays illuminate an area much smaller than the transverse sizes of the lens. The behavior of this lens is described by the so-called *lensmaker's formula* and expressed as [1]:

$$\frac{1}{f} = \frac{1}{d_o} + \frac{1}{d_i},$$ (C.1)

where f is the focal length, d_o is the distance between objects, and d_i is the distance between images. Figure C.1 shows the parameters for the study of ray tracing by a thin convergent lens, see C.1(a), and a thin divergent lens, see C.1(b).

Although equation (C.1) is quite useful, the general case is called a *thick* lens. In such a case, the lensmaker formula becomes [1]:

$$\frac{1}{f} = (n - 1)\left(\frac{1}{R_1} - \frac{1}{R_2}\right).$$ (C.2)

doi:10.1088/978-0-7503-4836-2ch10   C-1   © IOP Publishing Ltd 2025. All rights,

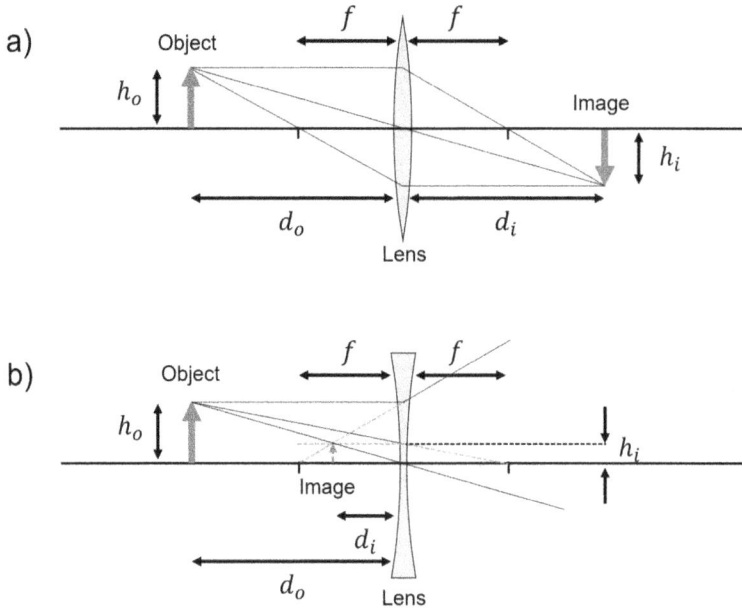

Figure C.1. Ray tracing and image formation in a thin convergent (a) and a thin divergent (b) lens.

Equation (C.2) relates the focal distance, f, the refractive index of the lens, n, and the surface curvature of the front, R_1, and back, R_2, faces of the lens. Figure C.1(c) shows the schematic representation of a thick convergent lens.

C.2.1 Convergent and divergent lenses

A convergent lens, also known as a concave lens, focuses incoming light rays into a point in front of the lens[1]. The divergent lens, also known as convex lens, diverts the light rays generating a focal point behind the lens. Figure C.1 shows the difference between these lenses.

C.2.2 Image formation

A lens forms an image by the intersection of light rays. It is usual to use three rays: one ray that runs parallel to the optical axis, one crossing the center of the lens, and one passing by the focal position and leaving the lens parallel to the optical axis [2], see figure C.1.

C.2.3 Lens magnification

Lens magnification refers to the change in size between *object* under study and the *image* generated by the lens, see figure C.1. The equation for the magnification, m,

---

[1] It is usual to say the front side of a lens is to the right in a scheme such as the one shown in figure C.1.

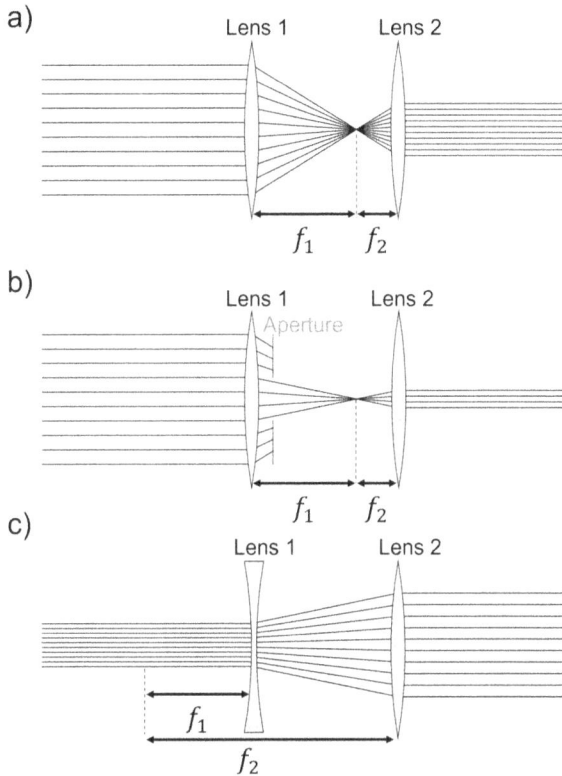

Figure C.2. (a) Beam reducer using a pair of lenses. (b) Beam reducer using two lenses and a physical aperture. (c) Beam expander using two lenses.

relates the height of the image (h_i) to the height of the object (h_o), as one would expect, and can be written as follows [1]:

$$m = -\frac{h_i}{h_o} \tag{C.3}$$

The parameters of equation (C.3) can be visualized in figure C.1[2].

C.2.4 Beam reducer and expander

An important application of lenses, in the context of dynamic diffraction, is the reduction and expansion of the size of light beams, while keeping them collimated. There are two ways to reduce the size of a beam. The first one is by using two lenses with different focal lengths, with the second lens located at a distance $f_1 + f_2$ in front of the first lens; figure C.2(a) shows this method. The second method uses the same scheme and adds a physical aperture, as shown in figure C.2(b).

_____

[2] It is important to mention that in the case of the divergent lens, figure C.1(b) the image is located behind the lens. In such a case, we say that this image is a *virtual* image.

A beam expander, on the other hand, can be built using one diverging lens followed by a converging lens located at a distance $f_2 - f_1$; figure C.2(c) shows the scheme of a beam expander.

C.3 Mirrors

Mirrors are reflective surfaces that allow the control of light beams by redirecting them [1]. When light falls on a mirror, it follows the law of reflection, i.e., the angle of incidence (θ_{inc}) is equal to the angle of reflection (θ_{ref}). Mathematically, this can be expressed as [1]:

$$\theta_{inc} = \theta_{ref}. \tag{C.4}$$

The most common types of mirrors are flat and curved mirrors.

The ray tracing of mirrors follows a procedure similar to that of lenses. It means, it is considered one ray parallel to the optical axis and its reflection, a second ray reflecting at the center of the mirror, and a third ray passing by the focal position, if there is any, and reflecting parallel to the optical axis. Figure C.3 shows the ray tracing of three types of mirrors that are discussed below.

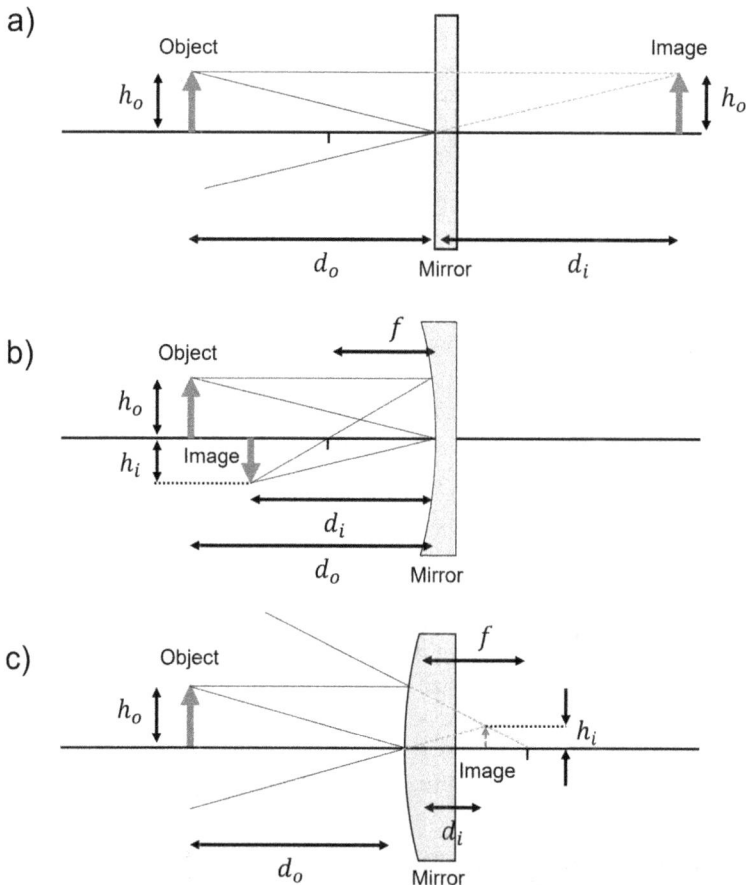

Figure C.3. Ray tracing and image formation in a concave (a) and a convex (b) mirror.

C.3.1 Plane mirrors

A plane mirror is a flat surface that reflects light rays without altering their size or shape. The image formed in a plane mirror is behind the mirror at the same distance as the object and upright [2]; figure C.3(a) shows the ray tracing in a flat mirror.

C.3.2 Curved mirrors

Concave and *convex* mirrors are also known as *curved* mirrors and can have any conic shape depending on the application. However, for the sake of simplicity, we only discussed spherical mirrors in this section. Spherical mirrors have the special property of having their focal point at half the radius of the sphere that generates the curved surface.

C.3.3 Concave mirror

A concave mirror curves inward, and its reflective surface is on the inner side; see figure C.3(a). The equation for the focal position of concave mirrors is given by [1]:

$$\frac{1}{f} = \frac{1}{d_o} + \frac{1}{d_i}, \tag{C.5}$$

with f the focal distance, d_o the distance to the object and d_i the distance to the image. All these are measured from the center of the reflecting surface.

It is important to note that concave mirrors can generate real and virtual images depending on the position of the object. Figure C.3(a) shows the case of the formation of a real image [2].

C.3.4 Convex mirror

A convex mirror curves outward, and its reflective surface is on the outer side. Convex mirrors always form virtual images with sizes smaller than the actual object [2]. The mirror equation for convex mirrors is [1]:

$$\frac{1}{f} = \frac{1}{d_o} - \frac{1}{d_i}, \tag{C.6}$$

with f, d_o, and d_i measured in the same way as with the concave mirror.

C.3.5 Mirror magnification

As in the case of lenses, curved mirrors can generate magnification. The magnification (m) of an image in a mirror is defined as the ratio of the height of the image (h_i) to the height of the object (h_o) [1]:

$$m = \frac{h_i}{h_o}. \tag{C.7}$$

C.4 Polarizers

As we have seen before, thermal sources, as discussed in appendix A, typically generate unpolarized light. However, it is possible to polarize such light using scattering or polarizers.

Polarizers are devices, usually made of strained polymers, that select a specific direction of oscillation of an incident light field. Polarizers can be found as linear, quarter-wave, and half-wave polarizers.

Linear polarizers generate linearly polarized light when the incoming light is unpolarized or control the output intensity when the incoming light is linearly polarized, according to Malus's law[3]. Quarter-wave plates generate circularly polarized light when linearly polarized light is used and its electric field is at 45° with respect to the fast-axis of the quarter-wave polarizer. A half-wave plate rotates the polarization of linearly polarized light and keeps it linearly polarized.

Figure C.4 shows the selection rules of common polarizers.

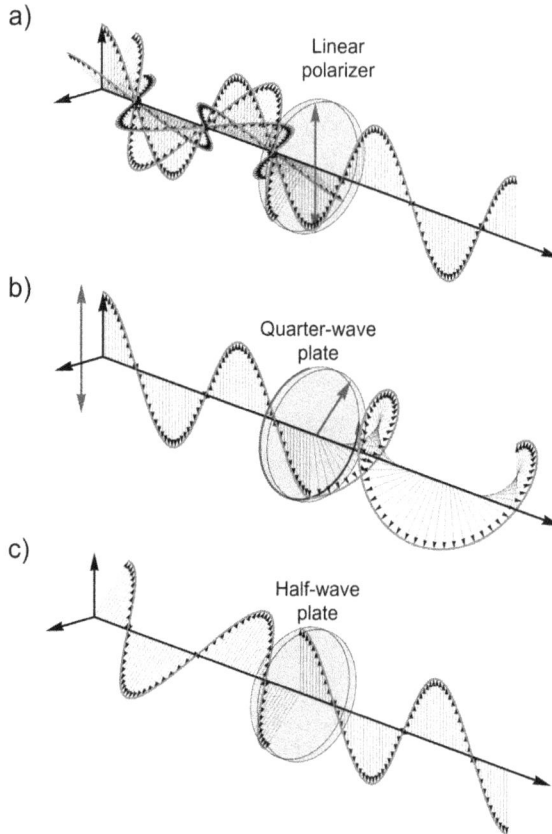

Figure C.4. Different use of polarizers. (a) Linear polarizer converting unpolarized light. (b) Linearly polarized light converted to circularly polarized light. (c) Rotation of linearly polarized light.

---

[3] Malus's Law states that the output intensity depends on the angle, θ, between the electric field and the polarizer axis according to $I = I_0 \cdot \cos^2(\theta)$; here I_0 is the incoming intensity.

References

[1] Hecht E 2002 *Optics* 4th edn (Boston, MA: Addison Wesley)
[2] Chandra S and Sharma M K 2024 *A Textbook of Optics* (Berlin: Springer)

IOP Publishing

Optical Interference and Dynamic Diffraction
Research methods for undergraduates
Jenny Magnes and Juan M Merlo-Ramírez

Appendix D

Safety precautions

D.1 Introduction

In this chapter, we will explore the most general safety precautions and rules to be followed inside a lab.

D.2 General lab safety

It is a good practice to follow the Occupational Safety and Health Administration (OSHA) guidelines for laboratory safety available elsewhere [1] and the Laboratory Safety Institute [2].

In addition, it is advised to implement a *safety buddy system* for the lab work. The safety buddy system consists in ensuring at least two people are present in the lab at all times when experiments are conducted. This ensures that if any of them are in trouble, there is another person to help, when possible, or notify the emergency services.

D.2.1 Electrical safety

It is important to keep in mind that most of the instruments used in an optics lab are electrically powered. In this sense, it is advised for the users to look for signs of wear in cables connected to the power outlet; damaged electrical cables can lead to electric shock or even electrocutions. Here, we make a compilation of the most basic safety suggestions for an optics laboratory, see below:

- Always inspect the equipment before use and report any issue to the person in charge.
- Look for damaged or broken cables. If any is found, replace it or contact a suitable person to do it.
- Use cables of different colors for each electrical polarity.
- Use adequate connectors when using external power supplies. Check the current and voltage requirements.

doi:10.1088/978-0-7503-4836-2ch11

- In the case of instruments with exposed electrodes, ensure they are out of reach and covered with the correct dielectric materials.
- Ensure all the outlets are grounded.
- When using equipment for the first time, ensure the voltage and current requirements are the correct ones.
- Use adequate signaling for high-voltage equipment.
- Use the adequate protection equipment when necessary.

It is advised for the reader to check the complete safety regulations at the Occupational Safety and Health Administration webpage [1], in particular Part 1910 subpart-S [3].

D.2.2 Chemical safety

Although it is not common for optics labs to deal with hazardous materials, it is important for the researches in the area to be aware of the hazards some common materials could create.

Below, there is a list of suggested preventive measures to improve safety in the lab when using chemicals:

- Label chemicals with the name of the material, the person responsible for the container, and the date of use.
- Label all the liquid containers legibly. Clear liquids in particular could become an issue.
- Use adequate protective equipment when using a chemical of any kind.
- Ensure that flammable solvents are stored in fireproof cabinets. Ethanol and isopropanol, commonly used for cleaning purposes, are particularly volatile.
- Never allow eating or drinking in a lab space.
- Store acids and bases separately. Store fuels and oxidizers separately [2].
- Do not allow food to be stored in chemical refrigerators [2].
- Provide fire extinguishers, safety showers, eye wash fountains, First aid kits, fire blankets and fume hoods in each lab and test monthly [2].
- Use warning signs to designate particular hazards [2].

It is also advised for the reader to read the Safety and Health Administration webpage [1], particularly Part 1910 subpart-H [4].

D.3 Laser safety

As mentioned before, nowadays most of the optics experiments are performed using lasers. As such, it is important to mention hazards and ways to prevent them. Below there is a compilation of the classification and hazards lasers could generate in a lab environment[1].

---

[1] It is advised for the reader to review the OSHA Laser Hazards description [5].

Table D.1. Laser classification by power output (continuous wave, visible range).

Laser class	Power range (CW)
Class 1	$\leqslant 0.39$ mW (visible)
Class 1M	Varies (large/divergent beam)
Class 2	$\leqslant 1$ mW (visible only)
Class 2M	$\leqslant 1$ mW (visible only)
Class 3R	$\leqslant 5$ mW
Class 3B	5–500 mW
Class 4	>500 mW

Table D.2. Laser classification based on potential eye and skin hazards.

Laser class	Eye hazard	Skin hazard
Class 1	None under normal operation	None
Class 1M	Hazardous with optical aids	None
Class 2	Low risk (blink reflex provides protection)	None
Class 2M	Hazardous with optical aids	None
Class 3R	Hazardous if directly viewed	Minimal risk
Class 3B	Dangerous (direct/diffuse exposure)	Possible burns at high powers
Class 4	Extremely dangerous (even diffuse reflections)	Can cause burns and fire

Lasers have been categorized according to their delivered power. A broadly accepted categorization is based on the American National Standard Institute (ANSI) Z136.1 [6]. This categorization is summarized in table D.1.

In appendix A, it was mentioned that lasers concentrate high power in small areas, i.e. lasers usually register larger intensities than other non-coherent sources, making them harmful if some conditions are met. Table D.2 summarizes the potential harms to the laser users[2].

It is important to note that even when table D.2 states that Class 1, 1M, and 2 are not harmful for the eyes, the use of optical aids, such as telescopes and microscopes, could make them a risk for the user.

As one would expect, the highest risk of injury during the use of lasers, and in general powerful light sources, is to the eyes of the user. As such, it is important to

---

[2] Diffuse reflection happens when light reflects out of a rough source. Contrarily, specular reflection happens when light is reflected from a smooth, mirror-like, surface.

Table D.3. Laser wavelength ranges and corresponding eye hazard regions.

Wavelength range (nm)	Region	Primary eye hazard
180–315	Far UV (UV-C)	Corneal damage (photokeratitis)
315–400	Near UV (UV-A/B)	Corneal and lens damage (e.g., cataracts)
400–700	Visible (VIS)	Retinal damage (focused light on retina)
700–1400	Near infrared (IR-A)	Retinal damage (penetrates to retina)
1400–3000	Mid infrared (IR-B)	Corneal and lens absorption; cataract risk
3000–10 000	Far infrared (IR-C)	Corneal and surface absorption (thermal burns)

understand how the wavelength of the light source could pose different effects on the eyes depending on the wavelength range[3]. Table D.3 is a summary of the possible effects different wavelength ranges could have on the user's eyes[4].

As mentioned before, the eyes are the organs at highest risk with the use of lasers. As such, there are several ways to protect the user eyes, among them, safety goggles, confined beams, and even electric interlocks that turn off the lasers is a specific condition is broken. This is mostly advised for lasers class 3B and 4.

Finally, good practices are advised when working in spaces where the space is shared. Below, there is a list of some of these good practices:

- Avoid desks and chairs at the same height than the optical tables to avoid undesired laser exposures.
- Contain the laser beam to the boundaries of the experimental setup.
- Create exclusion zones, clearly marked, when laser category 3B and 4 are in use.
- Use interlocks for lasers category 3B and 4.
- Use internationally recognized warning signs. Figure D.1 shows a set of these signs.

D.3.1 Laser safety goggles

As mentioned above, eyes must be protected when using lasers in classes 3R, 3B and 4. In this sense, the simplest method is the use of *laser safety goggles*. These goggles are made of specially designed materials that absorb light in specific wavelength regions. In this sense, the *optical density* (OD) is a measure of the transmitted power through a material. It can be defined as follows [7]:

$$OD = \log_{10} H_0/\text{MPE}, \tag{D.1}$$

---

[3] This is assuming the user is not taking enough preventive measurements.

[4] It is obvious that these effects will be dependent of the power of the laser, the kind of exposure, and the time of exposure.

[5] is the highest level of laser or other optical radiation to which a person may be exposed without experiencing harmful biological effects to the skin or eyes [6].

Figure D.1. Laser warning signs according to OSHA [7].

Table D.4. Optical densities for protective eyewear for various laser types.

Laser type (power)	Wavelength (nm)	0.25 s	10 s	600 s	30 000 s (8 h)
XeCl (50 W)	308	—	6.2	8.0	9.7
XeFl (50 W)	351	—	4.8	6.6	8.3
Argon (1 W)	514	3.0	3.4	5.2	6.4
Krypton (1 W)	530	3.0	3.4	5.2	6.4
Krypton (1 W)	568	3.0	3.4	4.9	6.1
HeNe (0.005 W)	633	0.7	1.1	1.7	2.9
Krypton (1 W)	647	3.0	3.4	3.9	5.0
GaAs (0.05 W)	840	—	1.8	2.3	3.7
Nd:YAG (100 W)	1064	—	4.7	5.2	5.2
Nd:YAG (Q switched)	1064	—	4.5	5.0	5.4
Nd:YAG (50 W)	1330	—	4.4	4.9	4.9
CO_2 (1000 W)	10 600	—	6.2	8.0	9.7

where MPE is the maximum permissible exposure level[5] and H_0 is the anticipated worst-case exposure [7].

Table D.4 shows the required optical densities for different kinds of lasers and exposure times [7].

References

[1] Occupational Safety and Health Administration *Laboratory* Safety https://www.osha.gov/publications/bytopic/laboratory-safety

[2] Laboratory Safety Institute *Laboratory* Safety Institute Resources https://www.labsafety.org/resource

[3] Occupational Safety and Health Administration *Subpart* S-Electrical https://www.ecfr.gov/current/title-29/subtitle-B/chapter-XVII/part-1910/subpart-S

[4] Occupational Safety and Health Administration. *Subpart* H-Hazardous Materials https://www.ecfr.gov/current/title-29/subtitle-B/chapter-XVII/part-1910/subpart-H

[5] Occupational Safety and Health Administration *Hazard* Recognition https://www.osha.gov/laser-hazards/hazards

[6] American National Standard Institute *ANSI* Z136.1 https://blog.ansi.org/ansi/ansi-z136-1-2022-safe-use-of-lasers/.

[7] Occupational Safety and Health Administration *OSHA* Technical Manual (OTM) Section III: Chapter 6 https://www.osha.gov/otm/section-3-health-hazards/chapter-6

IOP Publishing

Optical Interference and Dynamic Diffraction
Research methods for undergraduates
Jenny Magnes and Juan M Merlo-Ramírez

Appendix E

Solutions

Solutions to chapter problems.

E.1 Chapter 1 solutions

E.2 Chapter 2 solutions

E.2.1 Practice problem 2.8.5

Given: a Mach–Zehnder interferometer using a laser source of wavelength λ.

Task: determine the optical path difference between the two arms of the interferometer.

Physics: The optical path difference between the two arms of the interferometer is

$$\Delta L = L_2 - L_1. \tag{E.1}$$

Solution: the correspondent phase difference is given by:

$$\delta = \frac{2\pi}{\lambda} \Delta L. \tag{E.2}$$

Furthermore, the intensity in this interferometer would be expressed as:

$$\boxed{I(\Delta L) = 2I_0 \cos^2\left(\frac{\delta}{2}\right),} \tag{E.3}$$

assuming the amplitude of the fields are the same at the second beam splitter.

E.3 Chapter 3 solutions

E.3.1 Practice problem 3.4.2

Given: square slit with width $2a$ illuminated by a source located at infinity. It is assumed that the diffraction pattern is measured at the far-field.

Task: Determine the diffraction pattern at the far-field.

Physics: the diffraction pattern can be calculated using equation (3.27). In this case, the selection of Cartesian coordinates is the best choice due to the geometry of the problem. In such a case, the diffracted field is:

$$E(P) = E_0 \int_{-a}^{a} \int_{-a}^{a} e^{-2\pi i (f_x X_A + f_y Y_A)} dX_A dY_A, \qquad (E.4)$$

where E_0 is used to simplify the amplitude of the wave. Using the same assumptions we employed in the example in section 3.4.1, we see that the diffraction pattern is the following:

$$E(P) = E_0 \, \text{sinc}\left(\frac{2ax}{\lambda z}\right) \text{sinc}\left(\frac{2ay}{\lambda z}\right). \qquad (E.5)$$

Solution: The intensity of the diffraction pattern is simply the module square of equation (E.5), and is written as follows:

$$\boxed{I(P) = I_0 \, \text{sinc}^2\left(\frac{2ax}{\lambda z}\right) \text{sinc}^2\left(\frac{2ay}{\lambda z}\right).} \qquad (E.6)$$

Figure E.1 shows the diffraction pattern generated by a square aperture.

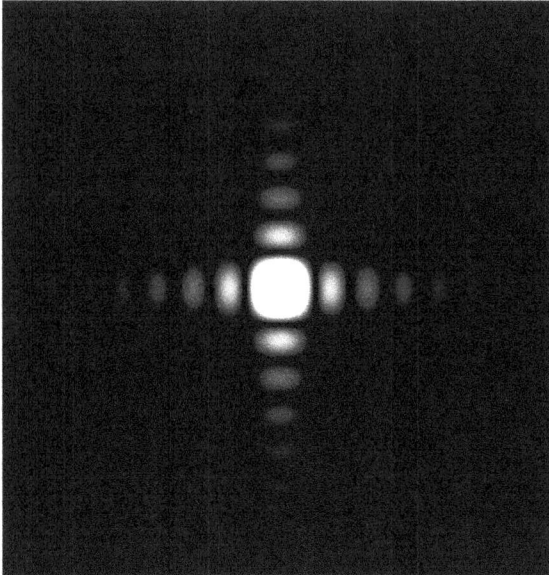

Figure E.1. Diffraction pattern generated by a square aperture. The wavelength used was 632 nm and the observation distance was 0.1 mm.

E.4 Chapter 4 solutions

E.4.1 Practice problem 4.3.2

Given: slit width a, slit separation b, distance from the slit to the screens d, distance from central maximum x_s

Task: $E = ?$; plot $I(x_s)$; are the results consistent with the location of previously approximated maxima at

$$x_s = \frac{m\lambda b}{d}. \tag{E.7}$$

Physics: The electric field E is equal to the Fourier transform of the aperture function $g(x)$ (diffracting object), in this case the double slit:

$$E = E_0 \int_{x=0}^{x=a} g(x)e^{ik_x x}dx. \tag{E.8}$$

The intensity I will then be proportional to the electric field E:

$$I \propto |E|^2 \tag{E.9}$$

Solution: We can use a combination of two single slits to represent the aperture function. We shift two aperture functions by $b/2$ in opposite directions and then add them to create a double slit—aperture function,

$$g(x) = \Pi_a(x - b/2) + \Pi_a(x + b/2). \tag{E.10}$$

The integral of the sum of two functions equals the sum of the integrals of each function:

$$E_0 \int_{x=0}^{x=a} (\Pi_a(x - b/2) + \Pi_a(x + b/2))e^{ik_x x}dx=$$
$$E_0 \int_{x=0}^{x=a} \Pi_a(x - b/2)e^{ik_x x}dx + E_0 \int_{x=0}^{x=a} \Pi_a(x - b/2)e^{ik_x x}dx. \tag{E.11}$$

Now, we can apply the shift theorem [1] to each integral:

$$E_0 \int_{x=0}^{x=a} \Pi_a(x \pm b/2)e^{ik_x x}dx = E_0 e^{\pm ik_x b/2} \int_{x=0}^{x=a} \Pi_a(x)e^{ik_x x}dx. \tag{E.12}$$

Combining equations (4.18), (E.11), and (E.12) the total electric field becomes:

$$E = E_0 e^{ik_x b/2} \int_{x=0}^{x=a} \Pi_a(x)e^{ik_x x}dx + E_0 e^{-ik_x b/2} \int_{x=0}^{x=a} \Pi_a(x)e^{ik_x x}dx$$
$$= E_0(e^{ik_x b/2} + e^{-ik_x b/2}) \int_{x=0}^{x=a} \Pi_a(x)e^{ik_x x}dx \tag{E.13}$$
$$= E_0 a(e^{ik_x b/2} + e^{-ik_x b/2})\operatorname{sinc}(k_x a/2).$$

Finally, Euler's identity $e^{ik_x b/2} + e^{-ik_x b/2}$ transforms into $2\cos(k_x b/2)$ so that the electric field reads

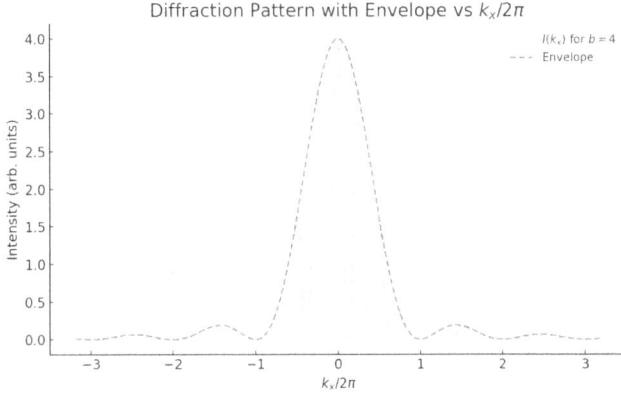

Figure E.2. The interference pattern due to a double slit with each slit width a 1/4 of the slit separation b. The envelope depicts the pattern typically shown in introductory physics textbooks without the finer interference structure.

$$\boxed{E = 2E_0 a \cos(k_x b/2)\,\mathrm{sinc}\,(k_x a/2)}. \tag{E.14}$$

The intensity I is proportional to the electric field magnitude squared:

$$\boxed{I = |E|^2 = 4E_0^2 a^2 \cos^2(k_x b/2)(\mathrm{sinc}\,(k_x a/2))^2}. \tag{E.15}$$

Using equation (E.15), figure E.2 shows that the first minimum of the envelope occurs around $k_x/2\pi \approx 1/a$. At this point, the sinc function equals zero and is consistent with equation (4.20) for the single slit.

The known result for constructive interference holds when $(m)\lambda \approx b\sin\theta$. For $m = 1$ so that:

$$\lambda \approx b\sin\theta \tag{E.16}$$

Also, $k_x = k\sin\theta$.

$$\lambda \approx b\frac{k_x}{k} \tag{E.17}$$

and with $k = \lambda/2\pi$,

$$k_x/2\pi \approx \frac{1}{b} \tag{E.18}$$

We had assumed that $a = 1$ and $b = 4$. Accordingly, we have a minimum at $1/a = 1$ and a maximum at $1/b = 1/4$. These results are in agreement with figure E.2.

E.5 Chapter 5 solutions

E.5.1 Practice problem 5.6.1

Centered DFT calculation for $\chi_{\Pi} = (1, 1, 0, 0)$

Let $N = 4$, so the input vector is:

$$\chi_\Pi = \begin{bmatrix} 1 & 1 & 0 & 0 \end{bmatrix}$$

Let $c = \frac{N-1}{2} = \frac{3}{2}$. The centered discrete Fourier transform is given by equation (5.7):

$$X_m = \frac{1}{N}\sum_{n=0}^{N-1}\chi_n \cdot e^{\frac{2\pi i}{N}(n-c)(m-c)}$$

Phase matrix ϕ

We use $e^{i\theta} = \cos(\theta) + i\sin(\theta)$:

$$\begin{bmatrix}
\cos\left(\frac{9\pi}{8}\right) + i\sin\left(\frac{9\pi}{8}\right) & \cos\left(\frac{3\pi}{8}\right) + i\sin\left(\frac{3\pi}{8}\right) & \cos\left(-\frac{3\pi}{8}\right) + i\sin\left(-\frac{3\pi}{8}\right) & \cos\left(-\frac{9\pi}{8}\right) + i\sin\left(-\frac{9\pi}{8}\right) \\
\cos\left(\frac{3\pi}{8}\right) + i\sin\left(\frac{3\pi}{8}\right) & \cos\left(\frac{\pi}{8}\right) + i\sin\left(\frac{\pi}{8}\right) & \cos\left(-\frac{\pi}{8}\right) + i\sin\left(-\frac{\pi}{8}\right) & \cos\left(-\frac{3\pi}{8}\right) + i\sin\left(-\frac{3\pi}{8}\right) \\
\cos\left(-\frac{3\pi}{8}\right) + i\sin\left(-\frac{3\pi}{8}\right) & \cos\left(-\frac{\pi}{8}\right) + i\sin\left(-\frac{\pi}{8}\right) & \cos\left(\frac{\pi}{8}\right) + i\sin\left(\frac{\pi}{8}\right) & \cos\left(\frac{3\pi}{8}\right) + i\sin\left(\frac{3\pi}{8}\right) \\
\cos\left(-\frac{9\pi}{8}\right) + i\sin\left(-\frac{9\pi}{8}\right) & \cos\left(-\frac{3\pi}{8}\right) + i\sin\left(-\frac{3\pi}{8}\right) & \cos\left(\frac{3\pi}{8}\right) + i\sin\left(\frac{3\pi}{8}\right) & \cos\left(\frac{9\pi}{8}\right) + i\sin\left(\frac{9\pi}{8}\right)
\end{bmatrix}$$

$$X = \phi \cdot \chi = \frac{1}{4}\begin{bmatrix}
-0.38 - 0.92i & 0.38 + 0.92i & 0.38 - 0.92i & -0.38 + 0.92i \\
0.38 + 0.92i & 0.92 + 0.38i & 0.92 - 0.38i & 0.38 - 0.92i \\
0.38 - 0.92i & 0.92 - 0.38i & 0.92 + 0.38i & 0.38 + 0.92i \\
-0.38 + 0.92i & 0.38 - 0.92i & 0.38 + 0.92i & -0.38 - 0.92i
\end{bmatrix}\begin{bmatrix} 1 \\ 1 \\ 0 \\ 0 \end{bmatrix}$$

Result

$$X = \begin{bmatrix} 0 \\ 0.33 + 0.33i \\ 0.33 - 0.33i \\ 0 \end{bmatrix}$$

Observations

X contains complex components where symmetry appears in X_0 and X_3 as they are complex conjugates: X_0 and X_3^*. Even though the complex and imaginary parts of X_0 and X_3 are equal in magnitude. While the aperture function does not represent a purely symmetric or asymmetric function, the symmetries are reflected in the DFT.

In this problem, we attempted to center the phase matrix. Of course, with the low number of elements, the system is still grainy. Still, the phase matrix shows similar symmetries, and the results in the example in section 5.5.1 also have conjugate symmetries as well as symmetries across the real and imaginary values.

E.5.2 Practice problem 5.10.2

1. Focus

Given:

A real-valued, symmetric 4×4 matrix χ:

$$\chi = \begin{bmatrix} 1 & 1 & 1 & 1 \\ 1 & 0 & 0 & 1 \\ 1 & 0 & 0 & 1 \\ 1 & 1 & 1 & 1 \end{bmatrix}$$

Discrete Fourier transform (DFT) is defined as:

$$\mathbf{X} = \frac{1}{N^2} F \chi F^\dagger,$$

where $N = 4$, F is the unitary DFT matrix, and $F^\dagger$ is its Hermitian transpose.

Objective:

- Compute the two-dimensional DFT of χ.
- Plot heat maps of:
    1. The real part of $\mathbf{X}$
    2. The imaginary part of $\mathbf{X}$
    3. The diffraction pattern (magnitude squared): $|\mathbf{X}|^2$

2. Ansatz

To obtain the 2D DFT of the given matrix, we use the property that the 2D DFT can be implemented using matrix multiplication:

$$\mathbf{X} = \frac{1}{N^2} F \chi F^\dagger$$

We choose the centered DFT convention for even dimensions.

$$F = \begin{bmatrix} \omega^{(-\frac{N}{2})(-\frac{N}{2})} & \cdots & \omega^{(-\frac{N}{2})(-1)} & \omega^{(-\frac{N}{2})(0)} & \cdots & \omega^{(-\frac{N}{2})(\frac{N}{2})} \\ \vdots & \ddots & \vdots & \vdots & \ddots & \vdots \\ \omega^{0\cdot(-\frac{N}{2})} & \cdots & \omega^{0\cdot(-1)} & \omega^{0\cdot0} & \cdots & \omega^{0\cdot(\frac{N}{2})} \\ \vdots & \ddots & \vdots & \vdots & \ddots & \vdots \\ \omega^{(\frac{N}{2})(-\frac{N}{2})} & \cdots & \omega^{(\frac{N}{2})(-1)} & \omega^{(\frac{N}{2})(0)} & \cdots & \omega^{(\frac{N}{2})(\frac{N}{2})} \end{bmatrix} \quad \text{where } \omega = e^{2\pi i/N}$$

For our 4×4 matrix, the F becomes:

$$F = \begin{bmatrix} \omega^4 & \omega^2 & \omega^0 & \omega^{-2} \\ \omega^2 & \omega^1 & \omega^0 & \omega^{-1} \\ \omega^0 & \omega^0 & \omega^0 & \omega^0 \\ \omega^{-2} & \omega^{-1} & \omega^0 & \omega^1 \end{bmatrix} = \begin{bmatrix} 1 & -1 & 1 & -1 \\ -1 & i & 1 & -i \\ 1 & 1 & 1 & 1 \\ -1 & -i & 1 & i \end{bmatrix}$$

and the hermitian conjugate:

$$F^\dagger = \begin{bmatrix} \omega^{-4} & \omega^{-2} & \omega^0 & \omega^2 \\ \omega^{-2} & \omega^{-1} & \omega^0 & \omega^1 \\ \omega^0 & \omega^0 & \omega^0 & \omega^0 \\ \omega^2 & \omega^1 & \omega^0 & \omega^{-1} \end{bmatrix} = \begin{bmatrix} 1 & -1 & 1 & -1 \\ -1 & -i & 1 & i \\ 1 & 1 & 1 & 1 \\ -1 & i & 1 & -i \end{bmatrix},$$

where we used Euler's identity $e^x = \cos x + i n x$. Notice that $FF^\dagger = 4\,\mathbb{1}$, hence the need for the normalization $\frac{1}{N^2}$.

$$\mathbf{X} = \frac{1}{N^2} \begin{bmatrix} 1 & -1 & 1 & -1 \\ -1 & i & 1 & -i \\ 1 & 1 & 1 & 1 \\ -1 & -i & 1 & i \end{bmatrix} \begin{bmatrix} 1 & 1 & 1 & 1 \\ 1 & 0 & 0 & 1 \\ 1 & 0 & 0 & 1 \\ 1 & 1 & 1 & 1 \end{bmatrix} \begin{bmatrix} 1 & -1 & 1 & -1 \\ -1 & -i & 1 & i \\ 1 & 1 & 1 & 1 \\ -1 & i & 1 & -i \end{bmatrix}$$

3. Solution
The resulting DFT is:

$$\mathbf{X} = \begin{bmatrix} 0 & 0 & 0 & 0 \\ 0 & -\dfrac{1}{8} & -\dfrac{1}{8} - \dfrac{1}{8}i & -\dfrac{1}{8}i \\ 0 & -\dfrac{1}{8} + \dfrac{1}{8}i & \dfrac{3}{4} & -\dfrac{1}{8} - \dfrac{1}{8}i \\ 0 & \dfrac{1}{8}i & -\dfrac{1}{8} + \dfrac{1}{8}i & -\dfrac{1}{8} \end{bmatrix}$$

We then extract:
- Re($\mathbf{X}$): real part
- Im($\mathbf{X}$): imaginary part
- $|\mathbf{X}|^2 = \text{Re}(\mathbf{X})^2 + \text{Im}(\mathbf{X})^2$: intensity (diffraction pattern)

The largest element is real, indicating the symmetric structure. It was impossible to center the phase matrix F so that asymmetric (imaginary) elements remained.

These components can be visualized as heat maps (figure E.3).

4. CHECK
- The DFT result is consistent with the symmetry of χ. The diffraction pattern shows a bright central peak and secondary lobes—typical of a framed square aperture.
- **Units:** Since χ is unitless, so is $\mathbf{X}$.
- Symmetry is preserved in $|\mathbf{X}|^2$.
- **Limiting behavior:** The heat maps are grainy and show asymmetry since the phase matrix cannot be centered in an even numbered structure. The small number of matrix elements worsen the issue as there is not enough over-sampling or sampling of the structure.

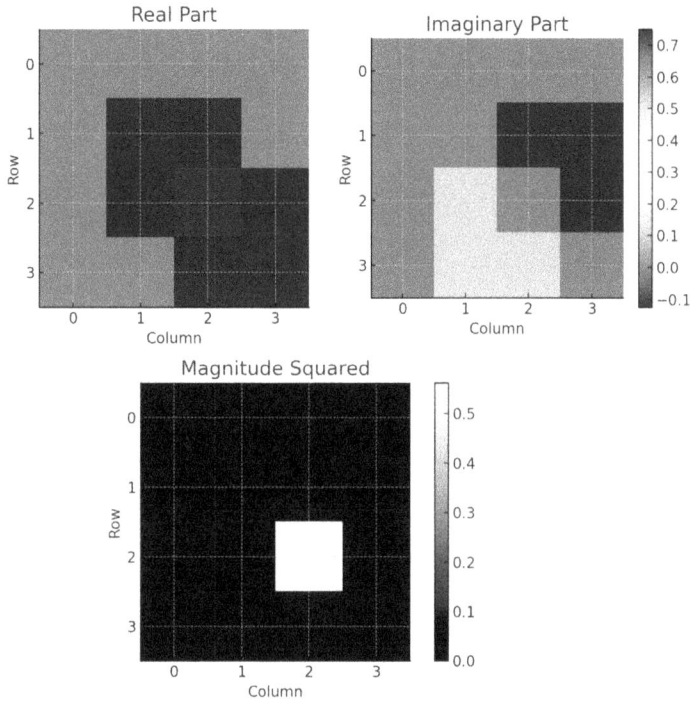

Figure E.3. Heatmap of the real, imaginary, and magnitude squared (diffraction pattern) of the DFT.

- In this case the structure consists of zeros rather than ones and the ones are the oversampling frame. This inverted structure still renders the same diffraction pattern as expected with the inverse structure; i.e., zeros and ones exchanged.

Solution verified and physically reasonable.

E.5.3 Practice problem 5.11.4

Oversampling ratio calculation

Consider the following binary matrix representing an object:

$$\chi = \begin{bmatrix} 1 & 1 & 1 & 1 \\ 1 & 0 & 0 & 1 \\ 1 & 0 & 0 & 1 \\ 1 & 1 & 1 & 1 \end{bmatrix}$$

We can use equation (5.28) and require that the oversampling ratio satisfies $\sigma > 2$, or we can consider the oversampling ratio of the entire area:

$$\sigma = \frac{\text{total area}}{\text{aperture area}} > 2$$

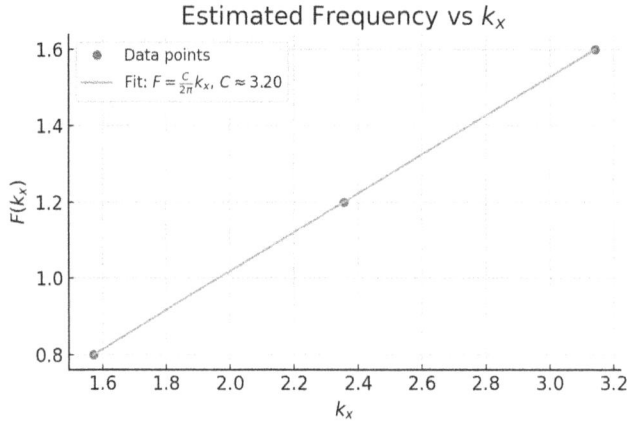

Figure E.4. Linear fit to data points extracted from figure E.4.

Let $N \times N$ be the size of the total matrix (support), and $M \times M$ the size of the object. In this case:

Consider the following matrix:

$$\chi = \begin{bmatrix} 1 & 1 & 1 & 1 \\ 1 & 0 & 0 & 1 \\ 1 & 0 & 0 & 1 \\ 1 & 1 & 1 & 1 \end{bmatrix}$$

This matrix is of size $N = 4$, but the central nonzero structure—the object of interest—is the 2×2 region of zeros (the hole), so:

$$M = 2$$

The oversampling ratio is defined as:

$$\sigma = \left(\frac{N}{M}\right)^2$$

Substituting in:

$$\sigma = \left(\frac{4}{2}\right)^2 = 4$$

Conclusion

Since $\sigma = 4 > 2$, this matrix **already satisfies** the oversampling condition for phase retrieval. No additional padding is necessary. Nevertheless, should we have to pad the matrix, we would pad it with ones instead of zeros.

E-9

$$\chi_{\text{padded}} = \begin{bmatrix} 1 & 1 & 1 & 1 & 1 & 1 & 1 & 1 \\ 1 & 1 & 1 & 1 & 1 & 1 & 1 & 1 \\ 1 & 1 & 1 & 1 & 1 & 1 & 1 & 1 \\ 1 & 1 & 1 & 0 & 0 & 1 & 1 & 1 \\ 1 & 1 & 1 & 0 & 0 & 1 & 1 & 1 \\ 1 & 1 & 1 & 1 & 1 & 1 & 1 & 1 \\ 1 & 1 & 1 & 1 & 1 & 1 & 1 & 1 \\ 1 & 1 & 1 & 1 & 1 & 1 & 1 & 1 \end{bmatrix}$$

E.6 Chapter 6 solutions

E.6.1 Practice problem 6.2.2

Estimation of slit expansion rate

We analyze the frequency of intensity oscillations as a function of spatial frequency k_x using the functional form:

$$F(k_x) = \frac{C \cdot k_x}{2\pi}$$

Solving for C, we get:

$$C = \frac{2\pi F}{k_x}$$

From the figure, we estimate the number of full oscillations within the interval $t \in [0, 5]$ for different values of k_x:

k_x	# Cycles	Frequency $F = \frac{\#\text{Cycles}}{5}$
$\frac{\pi}{2}$	4	0.8
$\frac{3\pi}{4}$	6	1.2
π	8	1.6

Using a linear fit to the data $F(k_x)$ versus k_x, we obtain:

$$F(k_x) = \left(\frac{C}{2\pi}\right)k_x \quad \Rightarrow \quad C \approx 3.20$$

Thus, the estimated expansion rate of the slit is:

$$\boxed{C \approx 3.20 \approx 2\pi}$$

Check: Comparing this result with the example plot in figure 6.2, we can see that the increased rate C induces a higher frequency for comparable values of k_x.

E.6.2 Practice problem 6.5.5

Given: Single slit with variable slit width $a(t) = Ct^2$ where $C = 1$.
  Task:
  (a) $I(kx, t) = ?$
  (b) Plot $I(k_x)$ for three values of t
  (c) Plot $I(t)$ for three values of k_x. Plot the fast Fourier transform (FFT) for all three values of kx.
  (d) What is the hallmark of a quadratic expansion?
  (e) How can you calculate the expansion constant C?
  Solve:
  (a) Start with equation (6.7) setting appropriate constants $E_0 = 1$ and inserting $a(t) = t^2$ so that

$$I(k_x, t) = t^4 \operatorname{sinc}^2\left(\frac{k_x t^2}{2}\right). \tag{E.19}$$

  (b) Figure E.5 shows the evolution of $I(k_x)$ as the slit narrows.
  (c) Adjusting equation (E.19) to account for a constant position k_x:

$$I(t) = \frac{4}{k_x^2}\sin^2\frac{k_x t^2}{2}. \tag{E.20}$$

Figure E.6 shows the time series at a single point in the diffraction pattern. The frequencies increase with both time and distance from the central maximum.

The FFT in figure E.7 shows nonlinear behavior and a continuous change in frequency.

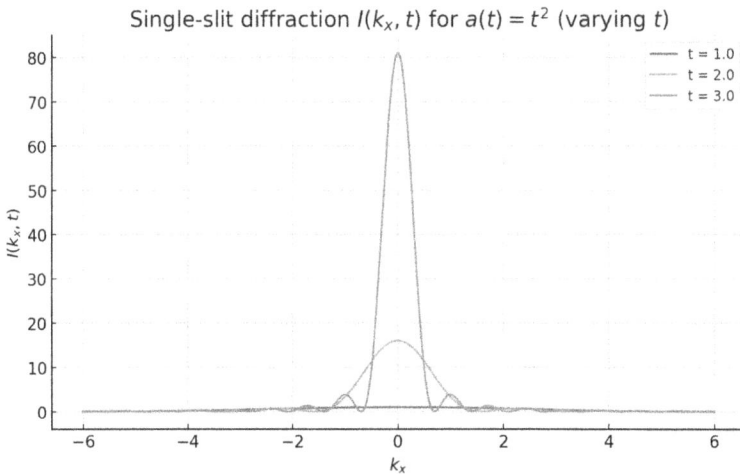

Figure E.5. The pattern is very low and wide when the slit is narrow. The pattern narrows and increases in amplitude as the slit widens.

$$I(t) = \frac{4}{k_x^2}\sin^2\left(\frac{k_x t^2}{2}\right), \quad t \in [0, 10]$$

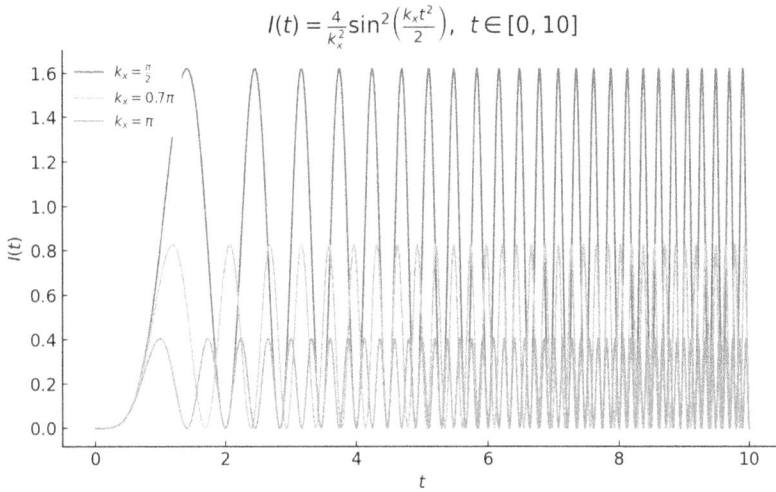

Figure E.6. The frequencies increase with time as the slit expands at an accelerating rate. Frequencies also increase with the distance k_x from the central maximum.

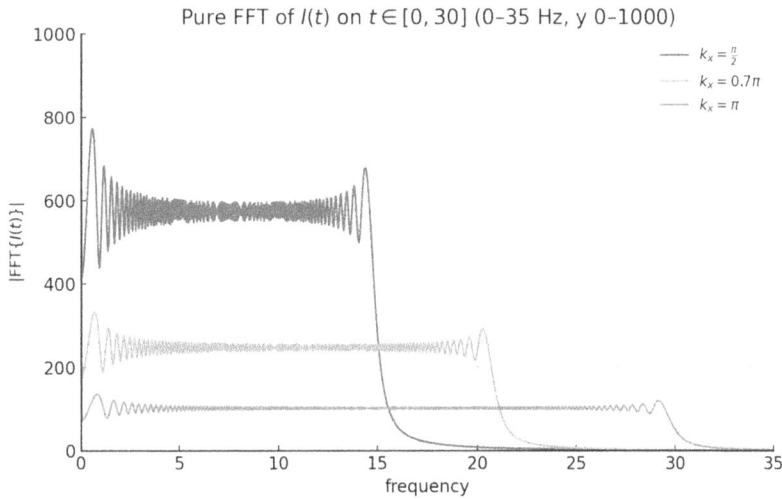

Figure E.7. FFT of a 30 second time series. Oscillation frequencies increase with position and time. The frequency range is governed by the observation time and position.

(d) The hallmark of the quadratic expansion is the continuously changing frequency and the period that is decreasing continuously with time.

(e) The expansion constant can be calculated in two ways: (i) using the time series from figure E.6 or (ii) using the FFT shown in figure E.7.

(i) Using figure E.6 the angular frequency $\omega(t)$ is the time derivative of the phase:

$$\omega(t) = \frac{\mathrm{d}\phi(t)}{\mathrm{d}t} = k_x C t = 2\pi f, \tag{E.21}$$

where T is the instantaneous period. We can rearrange equation (E.26) and plot the frequency $f = 1/T$ versus time t:

$$f = \frac{k_x C}{2\pi} t. \tag{E.22}$$

The slope m is then:

$$m = \frac{k_x C}{2\pi}, \tag{E.23}$$

so that

$$C = \frac{2\pi m}{k_x}. \tag{E.24}$$

(ii) We took the time series for 30 s. The highest frequency will be the cut off in the FFT. Equation (E.26) says that

$$f = \frac{k_x C t}{2\pi}, \tag{E.25}$$

so that:

$$C = \frac{2\pi f}{k_x t}. \tag{E.26}$$

We do have to remember; however, that the frequencies displayed in figures E.6 and E.7 are doubled by squaring the sine function. For that reason C is half of the calculated value.

E.6.3 Practice problem 7.7.1

A diffraction time series yields an FFT peak at 1.2 Hz and a secondary peak at 2.4 Hz. Interpret these results in terms of worm locomotion. How might the presence of additional peaks at 0.6 Hz and 1.8 Hz be explained?

FOCUS

Given: $I(t)$ $f = 1.2Hz$, $f_1 = 2.4Hz$, $f_2 = 0.6Hz$, and $f_3 = 1.8Hz$.

Objective: Determine the locomotion rhythm f_0 implied by the spectrum and explain the appearance of the extra peaks using the optics/physics of dynamic optical diffraction (DOD). Symbolically:

Infer f_0 from $\{0.6, 1.2, 1.8, 2.4\}$ Hz and interpret harmonic content.

Sketch (spectrum, for reference, figure E.8).

ANSATZ

Let f_0 denote the fundamental frequency of the body bend (biomechanical).

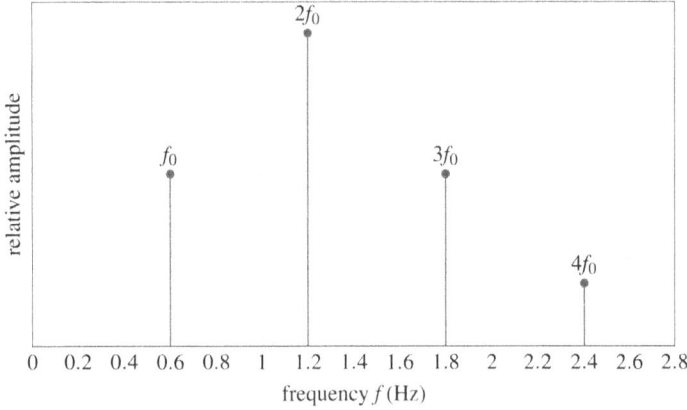

Figure E.8. Illustrative spectrum with peaks at 0.6, 1.2, 1.8, 2.4 Hz.

Physical principles. The detector measures *intensity*, not the field: $I(t) \propto |E(t)|^2$. A minimal 1D descriptor of the bend waveform is a scalar $u(t)$ with dominant periodicity f_0. The electric field oscillations can be described as a Fourier series u:

$$E(t) \approx c_0 + c_1 \sin(1 \cdot 2\pi f_0 t) + c_2 c_1 \sin(2 \cdot 2\pi f_0 t) + c_3 \sin(3 \cdot 2\pi f_0 t) + \cdots + c_n \sin(n \cdot 2\pi f_0 t), \tag{E.27}$$

where n is a positive integer.

Assumptions/approximations. (i) The worm executes approximately periodic bending with fundamental f_0. (ii) Sampling and SNR are sufficient so that spectral peaks represent physiology rather than aliasing or flicker. (iii) Weak non-sinusoidality/asymmetry allows finite c_1, c_3 (odd harmonics) while the intensity/symmetry often amplifies even harmonics via c_2.

Solution

Step 1: Model the bend.

Assuming that the lowest frequency is the fundamental frequency f_0, we can try various combinations of sines and cosines. For example,

$$I(t) = (3 \sin(2\pi \times 0.6t) + \sin(2 \cdot 2\pi \times 0.6t))^2, \tag{E.28}$$

results in an intensity pattern that matches the frequency distribution in the problem statement (figure E.9).

Step 2: Match observed peaks to integer multiples. The set $\{0.6, 1.2, 1.8, 2.4\}$ Hz fits

$$0.6 = f_0, \ 1.2 = 2f_0, \ 1.8 = 3f_0, \ 2.4 = 4f_0$$

with the *dominant* peak at 1.2 Hz explained by interference at $2f_0$.

Step 4: Interpret each reported feature.

- 1.2 Hz: (dominant): the even harmonic $2f_0$ emphasized by intensity detection and two-fold symmetry of body configurations.

E-14

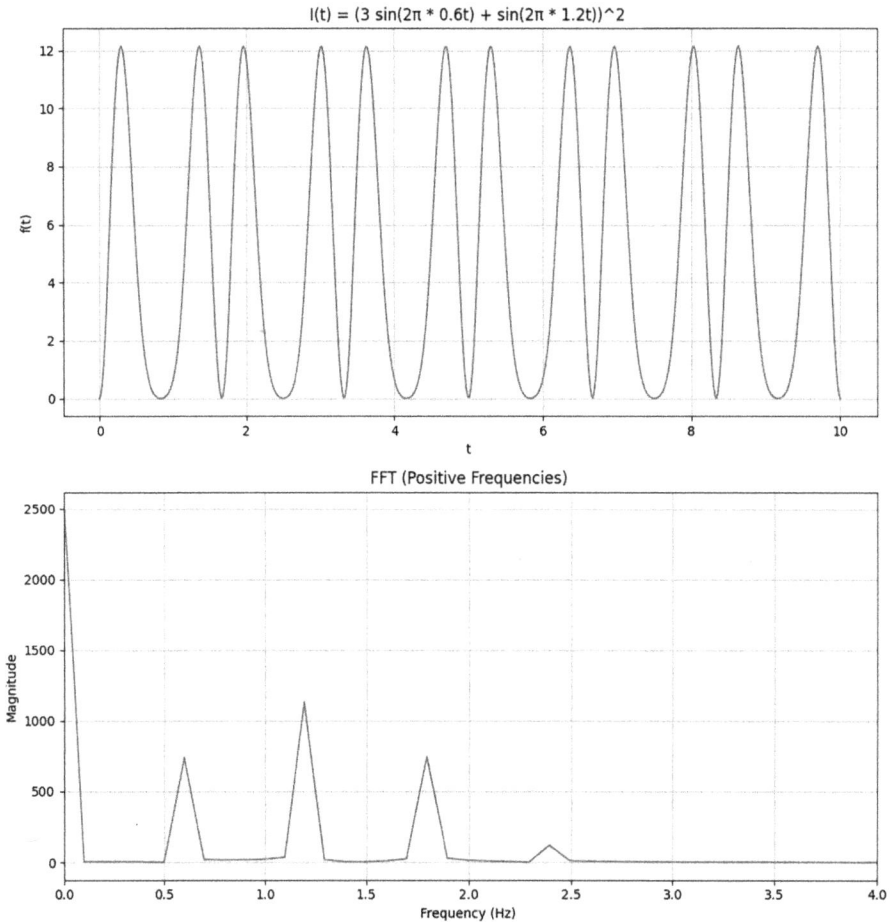

Figure E.9. (Top) Intensity $I(t)$ as a function of time. (Bottom) FFT to analyze the thrashing behavior of the worm. Showing peaks at $\{0.6, 1.2, 1.8, 2.4\}$ Hz.

- 2.4 Hz: next even harmonic $4f_0$ (higher-order nonlinear terms and waveform shape).
- 0.6 Hz: the biomechanical fundamental f_0, visible when symmetry is imperfect.
- 1.8 Hz: odd harmonic $3f_0$ arising from interference scattering ($c_3 \neq 0$).

Result. The locomotion rhythm is

$$\boxed{f_0 \approx 0.6 \text{ Hz}}$$

with intensity-dominated even harmonics at $2f_0$ and $4f_0$, and weaker odd harmonics at f_0 and $3f_0$.

Check

Units/order of magnitude. All frequencies are in Hz; intensities are dimensionless. A body-bend fundamental of ~ 0.6 Hz is plausible for small undulatory nematodes under typical conditions.

Reference

[1] James J F 2011 *A Student's Guide to Fourier Transforms: With Applications in Physics and Engineering* (Cambridge: Cambridge University Press)